P9-EDW-995

Global Environmental Diplomacy

Global Environmental Accord: Strategies for Sustainability and Institutional Innovation
Nazli Choucri, editor

Global Accord: Environmental Challenges and International Responses
Nazli Choucri, editor

Institutes for the Earth: Sources of Effective International Environmental Protection
Peter M. Haas, Robert O. Keohane, and Marc A. Levy, editors

Intentional Oil Pollution at Sea: Environmental Policy and Treaty Compliance
Ronald B. Mitchell

Institutes for Environmental Aid: Pitfalls and Promise
Robert O. Keohane and Marc A. Levy, editors

Global Governance: Drawing Insights from the Environmental Experience
Oran R. Young, editor

The Struggle for Accountability: The World Bank, NGOs, and Grassroots Movements
Jonathan A. Fox and L. David Brown, editors

The Implementation and Effectiveness of International Environmental Commitments: Theory and Practice
David G. Victor, Kal Raustiala, and Eugene Skolnikoff, editors

Global Environmental Diplomacy: Negotiating Environmental Agreements for the World, 1973–1992
Mostafa K. Tolba with Iwona Rummel-Bulska

Engaging Countries: Strengthening Compliance with International Environmental Accords
Edith Brown Weiss and Harold K. Jacobson, editors

The Greening of Sovereignty in World Politics
Karen T. Litfin, editor

Global Environmental Diplomacy

Negotiating Environmental Agreements for the World, 1973–1992

Mostafa K. Tolba
with
Iwona Rummel-Bulska

The MIT Press
Cambridge, Massachusetts
London, England

This book was set in Sabon by Northeastern Graphic Services, Inc.

Printed on recycled paper, and bound in the United States of America.

Library of Congress Cataloging-in-Publication Data

Tolba, Mostafa Kamal.
Global environmental diplomacy : negotiating environmental agreements for the World. 1973–1992 / Mostafa K. Tolba with Iwona Rummel-Bulska.
p. cm. — (Global environmental accord)
Includes bibliographical references and index.
1. Environmental policy—International cooperation—History.
ISBN 0-262-20113-5 (alk. paper)
2. Environmental law—History. 3. United Nations Environmental Programme—History. I. Rummel-Bulska, Iwona. II. Title.
III. Series: Global environmental accords.
GE170.T65 1998 98-18004
363.7'0526—dc21 CIP

Contents

Series Foreword

A new recognition of profound interconnections between social and natural systems is challenging conventional constructs as well as the policy predispositions informed by them. Our current intellect challenge is to develop the analytical and theoretical underpinnings crucial to our understanding of the relationships between the two systems. Our policy challenge is to identify and implement effective decision-making approaches to managing the global environment.

The Series on Global Environmental Accord adopts an integrated perspective on national, international, cross-border, and cross-jurisdictional problems, priorities, and purposes. It examines the sources and consequences of social transactions as these relate to environmental conditions and concerns. Our goal is to make a contribution to both the intellectual and the policy endeavors.

Foreword

The Montreal Protocol on Substances That Deplete the Ozone Layer is a major achievement in the struggle for the preservation of the global environment. Dr. Mostafa Tolba played a key role in formulating the Protocol and in transforming it into a political reality.

The depletion of stratospheric ozone by industrial chlorofluorocarbons (CFCs) is the global environmental issue that perhaps best illustrates the impact of human activities on the global environment. CFCs were developed as replacements for toxic refrigerants because of their chemical inertness and were hailed as miracle compounds. Ironically, it is this same property that allows them to reach the stratosphere, where high-energy ngtwhich in turn deplete the ozone layer that protects life on Earth from harmful solar ultraviolet radiation. CFCs are released predominately in the Northern Hemisphere, and yet the most striking effects of their release occur over the South Pole (as far as possible from the sources of contamination).

One very important lesson that we learned from the CFC-ozone depletion phenomenon is that society is capable of successfully addressing such global challenges. It illustrates how basic research, motivated by scientific curiosity to understand nature, can lead to practical results that benefit the society. The CFC problem is to a large extent under control, thanks to the Montreal Protocol. This Protocol not only provides an example of science in the service of public interest but sets a very important precedent that demonstrates how the different sectors of society (scientists, industry people, policymakers, and environmentalists) can work together and be very productive by functioning in a collaborative rather than in an adversarial mode. It emphasizes that global problems cannot be solved without the active participation of all the countries of the world (developing and developed). It has also established a new way of addressing environmental problems by applying the "precautionary principle": the original agreement was negotiated on the basis of the CFC–ozone depletion theory (which predicts that human-made CFCs would deplete the stratospheric ozone layer) rather than on direct unambiguous observation of noticeable damage to the ozone layer from CFCs. Another important precedent established by the Protocol is the inclusion of procedures for periodic revisions of its terms: the international agreement was strengthened first in London, in 1990, and then in Copenhagen, in 1992, leading to a complete ban on industrialized countries' manufacture of CFCs by the end of 1995. And, of course, society still enjoys the benefits of refrigeration, air conditioning, plastic foams, and aerosol spray cans—now with new, CFC-free technologies.

My own experience with the CFC–ozone depletion phenomenon offers me hope that as we face new environmental challenges (such as global warming, increased tropospheric ozone, loss of biodiversity, deforestation, and degradation of land), society will again rely on scientific understanding to provide the foundation for responsible action. The quality of life of future generations will be based to a large extent on our ability to deal intelligently with global-scale environmental problems.

As demonstrated by the story of the Montreal Protocol, negotiations leading to international agreements require the imagination and skills of inspired leaders. Dr. Tolba's dedication and personal credibility were crucial factors in the success of the Montreal Protocol negotiations. He traveled the world convincing key decisionmakers of the need for prompt political action. In addition, he also worked very successfully for treaties to protect the Gulf of Aden, the Red Sea, and the Mediterranean Sea. His contributions provide an invaluable legacy that paves the way for the solution of environmental problems with international dimensions.

Mario J. Molina
Institute Professor
Massachusetts Institute of Technology

Preface

What is the context within which a government concludes that an environmental issue must be addressed through international negotiation, and that the time is ripe for either informal agreements or a binding treaty? During the negotiations, what role is played by scientific research reports, business and industry, the media, nongovernmental organizations, politicians, and the public? Whereas the technical and diplomatic details of the achievement of environmental agreements have been covered in a number of academic journal articles, books, and university theses, there has been, to our knowledge, no effort to examine the almost unnoticed gradual movement of governments from the stance of holding absolute sovereignty over their actions to a more flexible concept that leaves room for international cooperation.

Between 1974 and 1992, the United Nations Environmental Program (UNEP) brokered a series of major environmental treaties; one or both of the authors played an active role in the negotiation process of all of them, Dr. Tolba as the executive director of the UNEP from 1976 to 1992, and Dr. Rummel-Bulska as chief of the UNEP Environmental Law and Institutions Unit beginning in 1982. Perhaps the most significant of the agreements whose emergence is described here dealt with ozone depletion, the disposal of hazardous waste, and the threats to biodiversity. Although these make up only a small fraction of the existing international instruments (declarations, charters, principles and guidelines, agreements, treaties), and although we took no part in the negotiation of some very significant treaties with important environmental aspects (the Partial Test Ban Treaty, the London Dumping Convention and the United Nations Convention on the Law of the Sea), it was here that we became involved to the fullest extent, cajoling, convincing, offering compromises, even arguing for what we believed to be the right principles. We learned that what went on outside the formal sessions, behind closed doors, in corridors, over dinner or drinks, was an integral part of the process, and although no one can claim to know the content of all these encounters, we could often sense what had happened as we saw their results.

We would like to share these insights, and our personal experience, with as wide an audience as possible. The present volume is not intended primarily as a textbook nor as reference material for scholars (who are being amply served by others: see our list of suggested readings), but as information to help readers in several fields: the theory and practice of environmental negotiations, the legal aspects of international environmental agreements, and the development of international cooperation in the environmental field. We have tried to use simple language without, as far as we could judge, frustrating the facts of either science or the law, and have minimized notes. Wanting to reach as wide an audience as possible, we have tried to convey our personal experience of the legal proceedings, to inject the human element rather than to present dry summaries, and thus to fill a gap that governments, scholars, negotiators, and others, including ourselves, felt existed.

Dr. Mostafa K. Tolba and Dr. Iwona Rummel-Bulska
Cambridge, Massachusetts
August 1997

Global Environmental Diplomacy

Prologue

Stockholm and Beyond: Sustainable Development

Development and the Environment

Our concept of environmental responsibility as we approach the twenty-first century is based on tenets expressed more than a century ago. In his 1864 classic *Man and Nature*, George Perkins Marsh tells us, "Man can control the environment for good as well as ill"; "Wisdom lies in seeking to preserve the balance of nature"; and "The present generation has an obligation, above all, to secure the welfare of future generations." At the turn of the century, Mohandas K. Gandhi, asked if he would like a free India to become like Great Britain, replied, "Certainly not. If it took Britain half the resources of the globe to be what it is today, how many globes would India need?" Still more recently, in his 1964 *Essays of a Humanist*, William Huxley viewed man both as a part of nature and as essentially unique, an opinion shared by many scientists; and with her book *The Sea Around Us*, which made the general American public aware of the dangers of pesticides, Rachel Carson in 1962 launched the modern popular environmental movement.

Still, not everyone agreed on the necessity to couple development with environmental responsibility, a concept that has come to be called "sustainable development." Many developing countries were suspicious of "environment" as a public issue on a global scale. This was partly because of how it was then defined, solely in terms of pollution abatement and as a technical problem, a restrictive interpretation stemming from the preoccupations of industrialized societies. Some developing countries viewed the outcry over environmental degradation as merely a tactic of the rich countries to keep the poor ones from industrializing; some of their leaders said that if pollution meant having industry, they would welcome it wholeheartedly, but some, recognizing that their own societies would have to face increasing pollution problems, wanted to avoid the mistakes already committed by the developed nations. Many feared that their ambitions to develop would suffer if the industrialized countries diverted their foreign-aid funds to environmental-protection projects. If anyone was going to lose out through a concern for the environment, they argued, it should be the rich countries, who are largely responsible for pollution in the first place. They also feared the additional expense of imported technology and machines that would meet environmental requirements, as well as damage to their export trade caused by newly adopted regulations in the developed world.

What could have been an impasse was given direction when a panel of experts from both developed and developing countries met, first in 1971 in Founex, Switzerland, to clarify the issues, then in Stockholm, at the 1972 United Nations Conference on the Human Environment. The Founex panel agreed that the many serious environmental problems of what was called during the cold war "the third world" (and is now increasingly being referred to as "the South," to differentiate it from the industrialized "North") were very often the result of extreme poverty and a lack of economic and social development. It was there that water quality, housing standards, sanitation, and nutrition were lowest, disease most prevalent, and natural disasters most destructive; and there that the impact of a fast-growing population resulted in the destruction of vital resources: soil, vegetation, and wildlife. To compound these ills, problems already distressingly familiar to the richer countries would emerge, perhaps sooner than many realized, as the poorer countries' industry, agriculture, transport, and communications expanded.

The Stockholm Conference clarified the link between development and the environment and suggested an approach that would recognize the socioeconomic factors behind many environmental problems and cure the effects by treating the causes. This was in contrast to the earlier technocratic approach, with its heavy emphasis on technology, itself directly descended from the Utopian dreams of the Industrial Revolution. Importantly, the conference redefined the aims of development, making a high quality of life, rather than the endless acquisition of material possessions, the main criterion of success. In this view, defining *environment* as the dynamic stock of physical and social resources available at any given time for the satisfaction of human needs, and *development* as a universal process aimed at increasing and maintaining human well-being, it became evident that an already complex issue had become even more so, that environmental and development objectives are complementary, and that the environmental agenda must be expanded.

A New Kind of Development

Thus, with the Stockholm Conference, a search began for a new, more rounded concept of development related to the limits of the natural-resource base and in which environmental considerations play a central role while still allowing opportunities for human activities. Clearly, earlier patterns of development were not sustainable. Current patterns of production and consumption, based on waste, extravagance, and planned obsolescence, would have to be replaced by those based on the conservation and reuse of resources.

Although this new kind of development has implications for both rich and poor countries and will lead to new directions for both growth and development while incorporating the environmental dimension, it will doubtless take different forms in the industrialized and developing countries. The developing world, lacking the infrastructure and available resources to meet the needs and aspirations of its people, will continue to work toward material ends; the industrial world must be encouraged to value the nonphysical areas of development that represent the highest levels of human achievement: the arts and humanities, education, and cultural pursuits.

Paradoxically, as growth and development continue, the industrialized world will find it more difficult to achieve environmental goals than the developing world. Heavily committed to the technological status quo even though it may already be inefficient, some economies cannot easily make rapid changes of direction; such large economic, social, and political investments have been made in wasteful production patterns that it is hard to take advantage of technological innovations that conserve resources or are more labor intensive. An example is the production of the private automobile, which has in some countries taken precedence over the maintenance of railways and the public transportation system.

On the other hand, the poorer nations, on the threshold of modern development, are favored by advances in technology that permit them to advance without destroying their resource base. Current practices that favor small-scale power projects, better suited to their needs, cultures, and value systems, will enable development to proceed without the destruction of the future environmental base and may give these countries a competitive advantage yet unconsidered. Such practices include developing local power supplies in place of the giant hydroelectric projects formerly favored by world lending agencies.

The South's need for economic aid and appropriate technology transfer from the North, and the conflicts this entails, become evident in the negotiations described in the following chapters.

Further Studies of Sustainable Development

At the beginning of the 1970s, the prestigious Club of Rome released *Limits to Growth*, warning that the planet's ability to sustain us, our industries, and our agriculture was being jeopardized, and that what for a relatively small human population had seemed infinite was actually alarmingly finite. Later studies, including some by the Club itself, would find the crisis somewhat overstated, but the report served to awaken scientists, policymakers, and the public to the dangers of environmental destruction.

Initially assuming that all resources were limited, the Club of Rome pointed to the oil crisis of 1974, warning that oil, a nonrenewable resource, would eventually become scarce. They cited the world's declining fish populations as an example of how renewable resources were being exploited at the margin, and in many cases had already crossed the margin of increase, and called attention to the extinction of more species during the twentieth century than in the preceding million years. Resources that could have sustained a smaller population forever, they said, were being snuffed out. Illustrations of this thesis could be found in the African Sahel and in the Horn of Africa, where the advance of modern medicine and transportation had permitted such an explosion in the human population that the land could no longer feed its inhabitants—

where growth had ignored the limits of the environment and become destructive.

The Stockholm Conference somewhat modified the dire warnings of the Club of Rome report, as it was pointed out that the Earth still has an essentially infinite supply of sunlight, a resource still largely untapped by man, and that many of our economic activities can be altered to make them work with, instead of against, the biosphere. In 1974, at the Cocoyoc Symposium on Patterns of Resource Use, Environment and Development Strategies, jointly organized by the UNEP and the United Nations Conference on Trade and Development (UNCTAD), thinking progressed further.

At the Cocoyoc Symposium it was agreed that

1. economic and social factors were often the root cause of environmental degradation, as patterns of wealth, income distribution, and economic behavior—both within and between countries—impeded development and led to inequities;
2. meeting the basic needs of the world's population was a primary goal, and the needs of the poorest, though urgent, must be met without jeopardizing the planet's carrying capacity;
3. different nations placed widely differing demands on the biosphere, with the rich preempting and wasting many cheap natural resources, whereas the poor were often left with no option except to destroy them;
4. developing countries must rely on their own judgment, rather than following in the footsteps of the industrialized world;
5. the principal means of achieving both environmental and developmental goals is to find alternative patterns of lifestyle and development; and
6. this generation must not jeopardize the well-being of future generations by squandering the planet's limited resources.

The Montevideo Process, 1981

By 1980, environmental law and environmentally sound development had risen among the priorities of many countries, including some in the industrialized world that had set up offices and adopted legislation to protect the natural environment. Enough treaties, guidelines, and principles had been developed under UNEP sponsorship that its Governing

Council convened an ad hoc meeting of senior government officials to establish a framework, methods, and a program for the development and periodic review of environmental law, including global, regional, and national agreements. This meeting, in Montevideo, Uruguay, in 1981, elaborated a comprehensive program identifying three major areas ripe for international cooperation and eight others that were deemed to merit action. The first three were to deal with land-based sources of marine pollution, protection of the ozone layer, and the transport, handling, and disposal of toxic and hazardous waste. Less urgent, but serious, were problems of international cooperation in environmental emergencies, coastal zone management, soil conservation, transboundary air pollution, international trade in potentially harmful chemicals, protection from pollution of inland waterways, legal and administrative mechanisms for prevention and redress of pollution damage, and methods of environmental impact assessment. All these issues were grouped under the title "the Montevideo Programme."

Some of the Montevideo initiatives led to binding international agreements: the Vienna Convention for the Protection of the Ozone Layer and the subsequent Montreal Protocol; the plans of action and their legal agreements for the Zambezi River and Lake Chad Basins; the Regional Seas Conventions and their protocols; the Basel Convention on the Control of Transboundary Movement of Hazardous Wastes and Their Disposal; and a number of regional agreements on land-based sources of marine pollution, under various conventions for the management of regional seas. Others have been developed into international guidelines, goals, and principles: the Cairo Guidelines on the management of hazardous waste; Goals and Principles for Environmental Impact Assessment; guidelines for offshore mining and drilling; and the London Guidelines for trade in potentially harmful chemicals. Still others are in the process of development as legally binding protocols within the framework of existing conventions, such as the Protocol on Liability and Compensation, under the Basel Convention on hazardous wastes.

In the meetings cited above, which were attended by representatives of both the developed and the developing world, the salient fact was that since Stockholm there has been a globalization of environmental issues. The alarming scientific evidence for ozone depletion and for the transboundary movement of air pollution as well as the unsolved problem of hazardous waste generation and transport energized governments to seek a new kind of cooperation. Since the Montevideo meeting, threats to biological diversity and the implications of climate change and global warming have been added to the list of problems that must be approached through international cooperation.

Toward Sustainable Development

Concerns voiced at Stockholm and the subsequent international symposia have led to the evolution of the theory and practice now known as sustainable development. At its core is the requirement that current practices not undermine future living standards: present economic systems must maintain or improve the resource and environmental base, so that future generations will be able to live as well as or better than the present one. Sustainable development does not require the preservation of the current stock of natural resources or any particular mix of human, physical, or natural assets, nor does it place artificial limits on economic growth, provided that such growth is both economically and environmentally sustainable. The parameters of sustainable development are broad, encompassing fiscal policy, international trade, industrial strategies, technology applications, labor rights, living conditions, natural resource conservation, and pollution reduction—in other words, all the components of development.

Although we have understood the link between environment and development for some decades, we are still at the starting gate of policy implementation. Economic priorities and ecological goals, though the subject of much rhetoric, have seen very little action. Since Stockholm a number of issues that were at that time considered regional—acid rain, pollution of shared fresh water resources, and drift net fishing, among others—have been recognized as international problems. Regional regimes and global accords, as well as nonbinding guidelines and principles, have begun to form a body of legal instruments constituting a new area of international law: international environmental law. In the process a new form of multilateral diplomacy, environmental diplomacy, has emerged.

In 1987, the World Commission on Environment and Development pointed out that most of the new wealth created in the 1980s had emerged from the nonpolluting industries: information and data processing and the service sector. In contrast, let us examine the agrochemical industry, which may be the prime example of short-term gains giving way to long-term losses, and which eloquently represents the technological approach now being questioned worldwide.

The "green revolution" that was to have saved world agriculture in the 1960s was based on the notion that the unlimited application of the pure chemical components needed for plant growth would result in an unlimited production of foodstuffs. Unfortunately, as much of the developing world's farmers (and some of the greatest food producers worldwide) have begun to realize, the long-term effect of this technological solution to the world's hunger problems is to increase, not reduce, the cost of production, add to environmental stress, and lead to the demise of the small farmer.

Food does not grow exclusively, mostly, or even largely because of man's efforts. Our efforts succeed because of the natural properties of the soil and the microorganisms that live in it, the biological absorption of nitrogen, and the energy of the sun. More pests are controlled by their natural predators than by chemical means. Yet modern agriculture has steadily increased the amount of pesticides and chemical fertilizers used, exhausting the soil's ability to support crops through natural biological regeneration; as crop yield drops, more fertilizer is applied, until the law of diminishing returns comes into play: the yield per unit of fertilizer falls. The crop output per cost unit of chemical input has decreased by half since 1960 and is still falling. Meanwhile, a toll is being exacted on both groundwater and surface waters as the fertilizer wasted by inefficient distribution techniques enters them and leads to over-enrichment and eutrophication.

The same pattern applies to pesticides: As pests become resistant, newer (and inevitably, more expensive) compounds must be devised; the economic cost, although great, is slight compared to the environmental cost, as toxins enter the food chain through surface application or groundwater pollution. In terms of everyday health effects on farm workers, the cost is great, and there remains the specter of industrial accidents during the manufacture of the products, such as the Bhopal disaster that killed more than 2,000 people and sickened up to a quarter of a million more.

Increasing environmental damage and declining economic output plague major agricultural areas throughout the world, but nowhere is this more apparent than in Latin America, where small farming is marked by an increasing number of bankruptcies, leading to an increase in urban migration. Each year, as agrochemicals become more expensive and less effectual, farmers and national treasuries must incur a higher debt; the need for ever more sophisticated chemicals forces farmers to turn increasingly to Northern manufacturers at a time when exchange rates are declining against the Western basket of currencies. Meanwhile the larger enterprises, giant farm corporations and agrochemical producers, having no interest in investing in or supporting alternative farming methods, continue to encourage dependence on their products. Even bilateral agencies are sometimes wary of promoting production techniques that might jeopardize national exports.

In terms of sustainable development, alternative methods of crop fertilization and pest control have shown that, although short-term production is often not as high as in the chemical-intensive farms, in the long term the net returns are the same and often higher, despite the small scale on which such so-called organic farms tend to operate. The use of natural predators to attack pests and the return to the soil of nutrients in the form of composted manure and plant matter produces no toxic waste, and less wasteful methods of water distribution reduce the strain on diminishing aquifers.

In 1983 the UNEP Governing Council called for two reports on environmental protection as it relates to the process of development. The reports appeared in 1987, when the World Commission on Environment and Development released *Our Common Future* and the UNEP published *Environmental Perspective to the Year 2000 and Beyond*. The reports are complementary; both elaborate environmental strategies for agriculture, health, transport, urban planning, and other key economic sectors.

Of course the road to sustainable development will not be smooth. It will demand a series of reforms to confront failed economic policies and instigate new structural adjustment programs. Although these reforms will in the long term alleviate poverty, meet basic human needs, and end economic conditions that promote environmental degradation, it will require imagination, determination, and courage to oust the rigidly unfair, protectionist international economic order, commodity price volatility, crippling debt, and chronic poverty that currently have a stranglehold on the South. Governments will need to set guidelines for themselves and for private enterprise that provide real, short-term material incentives for sustainable development, persevering in the face of the free marketeers who will be dismayed by such an idea.

But the father of capitalism, Adam Smith, did not suggest that private enterprise should work to the detriment of society as a whole. Quite the opposite: In his market system the profit of every individual reflected on the wider community; he would have been appalled by a system that knowingly undermined its own future. And had he been present to witness the extraordinary changes that have come about since his day, he would urge us to accept the principles of sustainable development.

1

The Evolution of International Environmental Law

For as long as humans have been altering their environment, they have been polluting it, and most of the time depleting its natural resources. Our search for better living conditions, longevity, wealth, and the comforts of life have through the millennia led to overexploitation of the planet and to a growing awareness that technological advances do not come without cost. Although it has been only in the last few decades that nations have begun to recognize their interdependence, some sense of neighborly responsibility is indicated as early as the 1273 English statute against air pollution and the royal proclamation of 1307 banning the use of coal in furnaces. At the level of customary law, particularly in the Western legal systems based on Roman law, the laws of trespass, nuisance, negligence, and liability have been combined with property laws to achieve some degree of control over resource management and pollution.

In the seventeenth century the Dutch jurist Grotius wrote the first and perhaps the greatest scholarly text on international law, *Mare Liberum* (Freedom of the Sea). He believed that the oceans beyond the reach of the shore should be free to all to use as they wished, subject only to the most basic constraints of civilized behavior; the question of how far a nation's direct reach could go would vex scholars, statesmen, and soldiers for hundreds of years. For a long while the rule was that a state controlled any water over which it could fire a cannonball from land. Originally this distance was set at one league, then three nautical miles, then twelve. Technical advances now permit nations to apply a 200-mile rule.

The issue of sovereignty has also yielded to the growth of modern technology and our power to affect our neighbors' environments. The underlying notion that each country's internal affairs are solely its business was established in 1648 at the Peace of Westphalia, which ended the Thirty Years' War. This formula led to the advent of nationalism and to the development of the state system during the eighteenth and nineteenth centuries—and ultimately to two world wars. Meanwhile, largely unnoticed, a challenge to the sovereign order was emerging as the Industrial Revolution eroded what one statesman called the "splendid isolation" of nations. In his famous letters on British imperialism in India, Karl Marx considered the impact of technological change on the international system; others followed, and by the end of the nineteenth century it was recognized that the premise of Westphalia was suspect: it was no longer clear that what one state did within its borders would have no material impact on other states.

The first attempts to form multilateral environmental agreements followed this recognition. Toward the end of the nineteenth century the riparian states of the Rhine basin entered into an agreement to protect the spawning grounds of the North Atlantic salmon, and in the interwar period of the 1920s and 1930s the League of Nations attempted to bring about a treaty on the control of marine pollution caused by ships. Although the latter effort failed, following the Second World War a number of international conventions on conservation were concluded. Mainly limited to the protection of migratory birds and wildlife, these conventions were ineffective because of sparse ratification and poor adherence. An example is the failure of the 1946 International Convention for the Regulation of Whaling, which was unable to slow the inroads of the whaling industry into a rapidly shrinking resource.

The technological feats accomplished during the Second World War on the one hand led to tremendous economic growth, increased life expectancy, and a population explosion; on the other, they fostered a deep unease as the world saw the destructive possibilities of atomic energy and witnessed a series of environmental disasters. In Japan mercury-bearing factory effluents in fish were identified as the cause of the deadly Minamata disease; it was announced that the blue whale, the largest animal on earth, was being hunted into extinction; in 1967 the tanker *Torrey Canyon* spilled thirty million gallons of crude oil off the southern coast of England; and in 1969 the uncontrollable eruption of an oil well being drilled in the rich waters of California's Santa Barbara Channel killed millions of fish, birds, and other marine life. Media coverage of rescuers' efforts to save the affected birds helped to focus public attention on ecological issues at a time when ecologists' proclamations about the fragility of the earth, the finite nature of its resources, and the consequences of their continued exploitation were being heard throughout the industrialized world. A number of ecological movements and associations sprang up at the local, national, regional, and international levels. Their concerns were addressed by a number of global and regional conferences initiated by the United Nations and its specialized agencies. These included the Biosphere Conference (officially, the Intergovernmental Conference of Experts to Consider the Scientific Basis for the Rational Use and Conservation of the Resources of the Biosphere), convened in Paris in 1968 by the United Nations Educational, Scientific and Cultural Organization (UNESCO); the Technical Conference on Marine Pollution and Its Effects on Living Resources and Fishing, which met in Rome in 1970 under the auspices of the UN Food and Agriculture Organization (FAO); and the Stockholm Conference in 1972. In addition, a succession of regional and global conventions were concluded covering several areas of the environment, especially marine pollution and the protection of natural resources.

All these events were occurring against the background of the Cold War and its political, military, and social alignments; in hindsight it appears that the concept of absolute sovereignty became blurred as issues that would prove most subversive of it emerged from apparently innocuous negotiations—those regarding human rights and shared natural resources. Sensitized by memories of wartime atrocities, movements within the liberal democracies particularly condemned the abuse of human rights in the Soviet Union and South Africa; those nations, echoing the Westphalian past, defended their right to do as they pleased within their own borders. The field of shared resources would yield results first, as negotiators honed their skills on international environmental agreements.

For the past forty years the international community has been trying to define the issue of the global commons: the oceans, the atmosphere, and some of what lies in between. It is evident that one state's behavior can radically change the amount of resources available to other states—drift net fisheries of one country may devastate a fishing ground used by all—and in the realm of management of the atmosphere there is no problem that does not cross national frontiers: acid rain, greenhouse gases and ozone depletion are salient at this writing.

The global environment is complex, and those who attempt to formulate international environmental law must perforce be specialists, each dealing in depth with a limited range of problems. The laws they shape require delicate negotiations and skillful diplomacy to find compromise solutions that satisfy the different political and economic motives of the states they represent. Often they are able to identify the real solution to the problem at hand, but the hard political and economic facts force adoption of a compromise measure, a half-solution that will be only the first step toward a final solution that may be years away. This results in the flexible nature of environmental treaties, which allows them to respond to changes in political and economic situations and to scientific revelations. For the same reasons, if national laws are passed to enforce the international agreements, they are seldom vigorously enforced.

Of course, the slow development of environmental law also slows the attainment of environmental goals; it is nevertheless considered the most valuable tool in policy implementation, and government-sponsored legal experts still meet with those from other disciplines to seek legal solutions to environmental problems. Many governments, especially those from the developing world, continue to ask for the help of United Nations agencies and foreign donors in formulating their own regulatory machinery.

One of the most important features of international environmental law as it develops has been its challenge to the doctrine of consent: the tradition that an agreement between two parties is binding only on them and not on third parties. As the transboundary nature of environmental threats has become clearer, a growing number of agreements have attempted to lay down norms and to offer incentives for other states to sign on, or at least to commit themselves to a similar course of action. The Convention on the International Trade in Endangered Species of Wild Fauna and Flora (CITES) is an example. The parties to CITES may trade with nonparties only if the latter substantially conform to the terms of the convention. Although this seems to undermine the liberalization of trade hammered out during more than a decade (during the Uruguay round of GATT, the General Agreement on Tariffs and Trade), it is justified by the realization that the internal activities of one country may have a material effect on others. A clear paradigm is the fact that, although Norway produces insignificant amounts of sulfur dioxide and nitrogen oxide, it has had one of the worst acid rain problems in the world, due to the activities of Britain, Germany, and others. The modern reality, deriving perhaps from the general principles of international law, is that Britain and Germany, even in the absence of specific regulations, have an obligation to refrain from causing harm to others. Similar examples are Canada's concern about the acid rain caused by industry in the United States; concerns in the United States about ozone depletion due to Asia's use of chlorofluorocarbons; and the danger that Egyptian farmers will be swamped by rising sea levels as a result of carbon dioxide emissions in the United States.

The 1972 Stockholm Declaration touched on these challenges to the absolute sovereign ideal when, in Principle 21, it declared: "States have . . . the sovereign right to exploit their own resources, pursuant to their own environmental policies, and the responsibility to ensure that activities within their jurisdiction or control do not cause damage to the environment of other States or of areas beyond the limits of national jurisdiction." This same principle would later be enunciated in the Rio Declaration, which is nonbinding, and in a legally binding treaty, the Convention on Biological Diversity, which, brokered by the UNEP, was signed in Rio de Janeiro during the June 1992 United Nations Conference on Environment and Development (UNCED, the "Earth Summit"), and became effective at the end of 1993.

These agreements and the growing spirit of cooperation they seem to indicate have taken place against the background of issues of national sovereignty, national debt, technology transfer, financial resources, and the cost of regulating development, all of which vary widely between developing and industrialized nations, and even within traditional voting blocs, frequently constituting an ideological gulf of significant proportions in spite of the universal nature of the environmental threat. In the face of this common problem, the essential motivation of each state remains self interest.[1] Governments will negotiate only if they believe the problem can be solved no other way; the decision to negotiate is strongly influenced by economic considerations, trade elements, pressure groups, and public opinion, all of which may be influenced by an announcement of new scientific findings.

After it has been agreed that an environmental treaty is needed, the negotiation process itself presents further difficulties. As new developments in science, technology, or economics become known, views of the negotiators change, creating difficulties in drafting the proposed treaty so that it is both reactive and anticipatory as well as flexible enough to adapt to new developments. This makes it difficult to draft a treaty that will impose binding standards, as shown by the problems in negotiating the treaties on ozone, hazardous waste, climate change, and biodiversity.

During the negotiations the conflict between economic necessity and ecological responsibility often becomes apparent, as the explanation is heard that "important bureaucracies at home favor economic growth." Public pressure is acknowledged to force governments into collective action for environmental protection, but the level of public concern does not explain policy inconsistencies among nations such as Canada, Germany, the Netherlands, and the United States on sulfur emissions that cause acid rain. Though all of these countries have experienced high levels of public pressure on this issue, their policies differ. This has been analyzed on the basis of whether the country in question is a net importer or exporter of pollution: If a net exporter and strong domestic groups oppose environmental protection regulations, public pressure will not be enough to shape a proenvironment policy.[2]

A further problem, common to the processes of establishing all international law, is that the inherent delays mean it may be years before the treaty goes into effect. First the negotiators must agree on a framework; then more time is needed to negotiate protocols on issues covered only in general terms in the framework; and finally there are normally long delays in parliamentary ratification by the parties. In the case of environmental treaties, an additional factor, one probably specific to them, is that the issues are characterized by a high degree of factual uncertainty, which can easily be converted to political issues and used as delaying tactics. The cost of alternative policy responses may also add to the difficulties, as do problems of identifying which parties are entitled to act in the matter at hand.[3] Generally, multilateral treaties come into force three to twelve years after they have been agreed upon; the average is about five years. Remarkably—and a clear indication of their importance to the contracting parties and their citizenry—with few exceptions the environmental treaties concluded over the past decade and their amendments have gone into force in less than two years.

An issue that should be clarified in environmental negotiations is the role of nongovernmental organizations (NGOs). At first rather limited, NGOs' activities have expanded to gathering technical information, devising policies, and mobilizing public support. Though NGOs should be fully involved in the negotiation process and given the opportunity to express their views, they should be governed by procedural rules and their activities monitored, because they are often important in urging governments to act but sometimes seek maximum results rather than achievable provisions, to be followed in time by stronger ones, as the more conservative model suggests.

The traditional approach to international treaties has not proven effective in the case of environmental agreements. Negotiating first a framework convention, which sets out the principles to be observed in solving the problem, and then a protocol that includes actual commitments, may have been useful in the early stages of international cooperation in the field of environment, but it is no longer appropriate. This approach encourages a lengthy negotiation process, and the signing of a convention couched in generalities rather than specifics may allow a political leader to substitute the symbol of signing for a real commitment. There are other weaknesses in the traditional convention-protocol procedure.[4] Most international environmental agreements include only nominal monitoring and enforcement provisions, which are difficult to negotiate as they are, rightly or wrongly, seen to conflict with the principle of national sovereignty. And with a few notable exceptions, negotiations have failed to link the environment with issues such as underdevelopment, debt, and poverty.

A number of innovations have been devised to induce agreement on environmental issues among otherwise reluctant parties. These include selective incentives, differential obligations, regionalization, and promotion of overachievement by lead countries.

In the first technique, selective incentives, additional benefits are offered a party to the agreement as encouragement to participate in a program it would otherwise find unacceptable. These may include access to funding, natural resources, markets, or technology. Examples are the Convention for the Protection of the World Cultural and Natural Heritage, in which parties maintain sites on the World Heritage List and are thus eligible for financial assistance to support those sites; the Montreal Protocol, whose Montreal Fund pays the incremental costs incurred by signatory developing countries in adhering to the control measures stipulated; the UN Law of the Sea Convention, guaranteeing access of its parties to shared mineral resources; and CITES, which gives access to world markets for wildlife products in return for observing agreed-upon conservation standards. The Montreal Protocol, the Basel Convention, and the Convention on Biodiversity are recent examples of treaties containing provisions for technology transfer in return for participation; the Convention on Biological Diversity also contains provisions for access by parties to biological resources.

The second approach, which provides differential obligations for parties to agreements, acknowledges that treaty obligations must be adjusted to the circumstances of each party. The Montreal Protocol offers a ten-year grace period for implementation to developing countries with less than 0.3 kilograms per capita consumption of the substances it controls. The EC's 1988 Directive on the Limitation of Emissions of Certain Pollutants into the Air from Large Combustion Plants and the 1976 Bonn Convention for the Protection of the Rhine against Pollution by Chlorides both used the principle of differential obligations. It has also been used in arranging contributions to various trust funds established since the early 1970s to finance joint programs under CITES, the Mediterranean Convention, the Transboundary Air Pollution Convention, the Vienna Convention on the ozone layer and its Montreal Protocol, and the Basel Convention on hazardous wastes.

In the third approach, regionalization, regional regimes have been established in some areas of international environmental law, since countries of one region usually share the same problems. These include the 1974 Helsinki and Paris Conventions for the Baltic and the North Sea; the UNEP regional seas agreements covering the Mediterranean, the

Persian/Arabian Gulf, the Southeast Pacific, the Caribbean, and West and Central Africa; and treaties for management of shared fresh water resources such as the Zambezi River and Lake Chad basins. In January 1991 the Organization of African Unity (OAU) adopted the Bamako Convention on the Ban of the Import into Africa and the Control within Africa of Transboundary Movement of Hazardous and Radioactive Wastes, which aligned a regional with a global regime, the Basel Convention. The Bamako Convention follows the basic principles of the Basel Convention, but is more stringent. Significant financial and technical resources are required for its implementation; non-African, industrial states not bound by the Bamako Convention are therefore essential to the effectiveness of the regional agreement, a cooperation that can be achieved only when these states and the African states have become parties to the Basel Convention.

By achieving more than an environmental treaty requires, participating countries encourage others to reach the treaty's goals; this is the philosophy behind the fourth apprach, the promotion of overachievement by lead countries. Some environmental treaties, such as CITES, the Montreal Protocol of the Vienna Convention, and the Geneva Convention on Long-Range Transboundary Air Pollution and its Helsinki and Sofia Protocols, contain provisions recognizing this initiative.

There are a number of devices used in international environmental agreements to avoid the delays inherent in the traditional treaty process. These include provisional application, "soft-law" options, delegated lawmaking and supplemental provisions within the treaty-making amendments or adjustments binding on all signatories not specifically opposing them. The first device, provisional application, is a procedure recognized by the Vienna Convention on the Law of Treaties, and may be exemplified by the Geneva Convention on Long-Range Transboundary Air Pollution, whose signatories decided to "initiate, as soon as possible and on an interim basis, the provisional implementation of the convention," and to "carry out the obligations arising from the convention to the maximum extent possible pending its entry into force." This resulted in the establishment of an interim executive body, which held regular annual meetings and created subsidiary working groups well before the convention took effect in 1983. This was also the case with the Basel Convention of 1989, whose plenipotentiary conference also resolved "that until such time as the Convention comes into force, all states refrain from activities which are inconsistent with the objectives and purposes of the Convention." All states were called on to apply the provisions of the Basel Convention as soon as possible.

The second device employed to avoid implementation delays is soft-law options. The term "soft law" is used to distinguish informal agreements (such as codes of conduct, guidelines, and principles) from formal, legally binding agreements. Many such instruments have been formulated in the environmental field, covering shared natural resources, offshore mining and drilling, the exchange of information on chemicals in international trade, environmentally sound management of hazardous waste, marine pollution from land-based sources, weather modification, and environmental impact assessment, as well as the (nonbinding) forestry principles adopted by the UNCED. The great advantage of these agreements is that they do not require ratification and can be put into use immediately; this is, however, offset by their informality and lack of legal force. Soft-law agreements may eventually evolve into treaties. This was the case with the Cairo Guidelines and Principles for the Environmentally Sound Management of Hazardous Wastes, the basis of the Basel Convention.

In the case of treaties dealing with areas subject to frequent changes in technology and the rapid advance in scientific knowledge, a third device, the delegated lawmaking approach, permits an intergovernmental body to revise the treaties without the need for ratification by the parties. An example is the control measures of the Montreal Protocol on Substances That Deplete the Ozone Layer.

There is a consensus that global environmental protection is an appropriate realm of international law. As this topic has moved from the scientific world to that of high politics, governments have taken a more flexible approach to the question of national sovereignty. Environmental law has developed rapidly since its inception in the 1960s, with negotiations in the field being complicated by scientific uncertainties. This has led to the flexible, anticipatory nature of environmental agreements, and to the use of techniques that will enhance their adoption, ratification, and implementation.

Notes

1. Several analysts have identified two major forces influential throughout the negotiating process: the parties themselves and the perceived value of their interests, preoccupations, and constraints. See Albin 1992; Carnevale and Pruitt 1992; Carroll 1988; Cross 1978; Graham and Trotman 1992; Hurrel and Kingsbury 1992; Kremenyuk 1991, 1992; Lang 1992; Porter and Brown 1991; Sand 1992a,b,c; Sjostedt and Spector 1993; Spector 1992, 1993a, 1993b, 1993c; Susskind 1994.

2. See Boehmer-Christiansen 1992; Francioni and Scovazzi 1992; UNCED Conference 1991.

3. A typical example is the role of the European Community (EC), now the European Union (EU). Although the EU has clearly established itself as an actor on environmental problems, because of its structure, problems of competence arose during some negotiations, when member states also retained a degree of competence.

4. See Hurrel and Kingsbury 1992; Lang 1992; Porter and Brown 1991; Sand 1992; Spector 1992, 1993a, 1993b, 1993c; Susskind 1994.

2

Environmental Soft Law: Guidelines and Principles

When an environmental issue is ripe for quick international action, but governments are not prepared to enter into the treaty process, soft laws may be enacted. These are agreed-upon codes of conduct, guidelines, and principles that can be put into effect without a lengthy ratification process.

Negotiations of this kind began in 1976, with a text entitled *Principles of Conduct in the Field of the Environment for Guidance of States in the Conservation and Harmonious Utilization of Natural Resources Shared by Two or More States*, known as the Shared Natural Resources Guidelines. Legal experts representing their various governments met under the auspices of the UNEP and found the negotiations to be as difficult as if they were working out a binding treaty. When the issue was raised in the UNEP Governing Council, the head of the Brazilian delegation (Bernardo Brito, later deputy executive director of the UN World Food Program) led his delegates from the room in protest sparked by Brazil's conflict over the La Plata River, which it shared with Argentina.

During the period covered by this book, seven such nonbinding guidelines and principles were negotiated under the auspices of the UNEP,[1] setting a series of important precedents in environmental management. Their titles indicate how wary the participating governments were of committing themselves, in spite of the nonbinding nature of the instruments.

Shared Natural Resources

In response to a 1973 United Nations General Assembly resolution, a working group of legal experts was established that between 1976 and

1978 developed draft principles on shared natural resources. The principles were adopted in 1979, and all states were requested "to use the principles as guidelines and recommendations in the formulation of bilateral or multilateral conventions regarding natural resources shared by two or more States, on the basis of the principles of good faith and in the spirit of good neighborliness and in such a way as to enhance and not to affect adversely development and the interests of all countries and in particular of the developing countries." This succinctly encompasses a number of significant concepts: It reconfirms Principle 21 of the Stockholm Declaration; Principle 3 of the shared natural resources guidelines states that nations have the sovereign right to exploit their own resources as well as the responsibility not to injure states outside their jurisdiction; it is a reminder that the agreement contained only principles to be used as guidelines and recommendations by its parties in a spirit of good faith and good neighborliness; and it put special emphasis on the development and interests of developing countries while assuring that it would not adversely affect those of all countries.

These elements are still present in every environmental negotiation. Each state must as much as possible avoid adverse environmental effects beyond its jurisdiction in the use of a shared natural resource, especially when such use might affect the use of the resource by a sharing state, threaten the conservation of a shared renewable resource, or endanger the health of another state's populace. This principle should be interpreted as taking into account the practical abilities of sharing states to abide by it.

Principle 6 sets out the requirement that every sharing state notify others in advance of any plans to begin or change its use or conservation of the resource if this will significantly affect other states' environment, and upon request to enter into consultations regarding its plans, providing any additional information requested; if the state has not furnished such advance notice, it must upon request consult with states that may be affected.

In addition, Principle 9 states that states have an urgent duty to inform other states that may be affected by any emergency situation arising from use of a shared natural resource, or from sudden natural events related to the resource, that may cause harmful effects on their environment; Principle 13 requires that domestic environmental policy take into account the potential adverse environmental effects of use of shared natural resources, whether the effects were in their jurisdiction or outside it. These two principles constitute a significant erosion of the old rule of absolute sovereignty.

Provisions for Cooperation between States in Weather Modification

Weather modification guidelines adopted by the UNEP Governing Council in 1980 similarly required early notification of any planned practices that might affect other states. Environmental impact assessments appeared for the first time in this agreement, much predating the Governing Council's 1987 Environmental Impact Assessment. This requires that "states should ensure that an assessment is made of the environmental consequences of prospective weather modification activities under their jurisdiction or control which are likely to have an effect on areas outside their national jurisdiction, and either directly or through the World Meteorological Organization, make the results of such assessments available to all concerned states." Similarly, states were to notify potentially affected states and enter into "timely consultation" with them.

Offshore Mining and Drilling

In 1982, the UNEP adopted guidelines on offshore mining and drilling, formally titled *Conclusions of the Study of Legal Experts Concerning the Environment Related to Offshore Mining and Drilling within the Limits of National Jurisdiction*. These guidelines follow Principle 21 of the Stockholm Declaration and for the first time provide guidance on how to perform an environmental impact assessment, as well as defining elements of liability and compensation in cases of environmental damage.

The guidelines are an attempt to formalize management of offshore exploration and drilling for oil and other minerals so as to minimize pollution and other harmful results, and urge states to adopt laws regulating these activities. They include recommendations for assessment of environmental effects, advance notice measures, assignment of liability and payment of compensation for damages. The text of the guidelines is presented as appendix A to this chapter.

The Montreal Guidelines

In 1985 the UNEP Governing Council adopted the Montreal Guidelines for the Protection of the Marine Environment against Pollution from Land-Based Sources, which had taken almost two and a half years to negotiate. Following the pattern of earlier guidelines, they include elements from Principle 21 of the Stockholm Declaration and continue to move away from the notion of absolute sovereignty.

The Montreal Guidelines exemplify the use of nonbinding agreements to achieve cooperation in an area ripe for a legally binding global treaty.

Environmental Impact Assessment (EIA)

The goals and principles of environmental impact assessment were adopted by the UNEP in 1987. These guidelines codify the processes that, beginning with Principle 21 of the Stockholm Conference, had been developing throughout the history of the agreements described above. They would be used in the 1991 Convention on Environmental Impact Assessment (EIA) in a Transboundary Context, which was elaborated under the auspices of the United Nations Economic Commission for Europe (ECE) by twenty-six states and the EC, meeting in Finland. EIA is also mentioned in a large number of conventions, protocols, and similar instruments of international law.[2]

All these instruments differ in their terminology and in the obligation they impose on their parties to carry out the environmental impact assessments; some of the earlier ones do not expressly use the term, and others refer to "environmental assessment." The Convention on the Regulation of Antarctic Mineral Resource Activities speaks of the assessment of the possible impacts of such activities. Some of the instruments identify the minimal content of an EIA, but none specifies procedures or methods. As yet, there have been only a few cases in which rules have been made for procedural details or to name the activities for which an EIA should be carried out; exceptions are the 1987 Antarctic Treaty Consultative Meeting's recommendation for an EIA on human impact on the Antarctic, and provisions dealing with EIAs of seabed activities under the UN Convention on the Law of the Sea. The Action Plan for the Environmentally Sound Management of the Common Zambezi River System of May 1987 addresses environmental assessment as one of its four elements, recognizing the need for continuing systematic assessment of water management and quality and describing the tasks required, but it does not specify the administrative procedures to be followed. The United Nations Principles on Shared Natural Resources and the World Charter for Nature also included references to environmental assessment.

Environmental impact assessment is also mentioned in a number of bilateral agreements, for example, in the Agreement on Cooperation for the Protection and Improvement of the Environment in the Border Area concluded between the United States and Mexico in August 1983 and in force since February 1984.

Further legal instruments requiring an EIA for specific activities are being developed at this writing. They include, under the Kuwait Regional Convention, a draft protocol concerning marine pollution resulting from exploration and exploitation of the continental shelf; under the Cartagena Convention, a draft protocol concerning specially protected areas and wildlife; under the Antarctic Treaty, a draft protocol on environmental protection and an annex to the protocol laying down rules on the EIA of proposed activities in the area covered by the treaty; and the European Community EIA Directive adopted in 1985 by the Council of the European Community.

All the above shows a clear move away from the old concept of national sovereignty, but in spite of this, attempts to develop the guidelines into an international treaty have failed. The next step must be global legislation that will protect the environment and—even more important—will reduce conflicts and enhance peace by making each nation responsible for the damage its activities may inflict on its neighbors.

Management of Hazardous Waste

An example of how nonbinding guidelines and principles can help states deal with a common problem pending a later binding agreement is the Cairo Guidelines and Principles for the Environmentally Sound Management of Hazardous Wastes, adopted by the UNEP Governing Council in 1987, which became the basis for the Basel Convention on the

Transboundary Movement of Hazardous Wastes and their Disposal. (The negotiating process for the Basel Convention is described in chapter 7.) The Cairo Guidelines included four important principles governing the transfer of technology, capacity building, public access to information, and liability and compensation. The guidelines recognized developing countries' need for technical assistance from the industrialized world to ensure the environmentally sound management of hazardous waste, pointed out the need for the public to have access to all information concerning this activity, and established a requirement for national laws governing liability and compensation for damages in case of accident or mismanagement.

International Trade in Chemicals

In 1972 the United Nations Conference on the Human Environment requested that the UN, governments, and scientific and international bodies "develop plans for an International Registry of data on chemicals in the environment based on the environmental behavior of the most important man-made chemicals, together with their pathways from factory via utilization to ultimate disposal or recirculation." This request resulted in the establishment of the International Register of Potentially Toxic Chemicals (IRPTC) in 1976, a computerized data bank for the exchange of information on production and consumption of these substances, their use, treatment of poisoning, waste management and control of hazards posed by them.

In the late 1970s Kenya called the attention of the UNEP Governing Council to the indiscriminate trade in toxic chemicals, particularly that carried on in the South by companies in the North; in 1977 the council urged governments "to take steps to ensure that potentially harmful chemicals . . . which are unacceptable for domestic purposes in the exporting country, are not permitted to be exported without the knowledge and consent of appropriate authorities in the importing country." In 1978 the council further urged that such chemicals not be exported until their health and environmental effects had been tested and reported to the recipient countries, and called for adequate monitoring, evaluation, and protections to be instituted by both exporting and importing governments.

The Montevideo Programme, adopted by the Governing Council as its program of action in environmental law for 1981–1991, included the preparation of principles or guidelines on the exchange of information relating to the trade in potentially harmful chemicals, particularly in pesticides. An ad hoc working group of experts met in the Netherlands in March 1984 and submitted to the next Governing Council session a Provisional Notification Scheme for Banned and Severely Restricted Chemicals, which the council adopted, calling for it to go into effect as soon as possible and urging close cooperation with other UN bodies and specialized agencies in further elaboration of the draft guidelines. Among its provisions, the notification scheme called for a country that has acted to ban or severely restrict a chemical to notify other nations' authorities promptly and to provide the appropriate information at the time of the first export following such action.

In 1987 the UNEP Governing Council adopted the London Guidelines for the Exchange of Information on Chemicals in International Trade, which continued to be developed and two years later were adopted as amended. The 1987 guidelines were designed to complement existing instruments of other governments and intergovernmental organizations, and of the UN and its specialized agencies, particularly the FAO's code of conduct on the distribution and use of pesticides. The guidelines stated that in using chemicals, nations should abide by Principle 21 of the Stockholm Declaration, and flagged a potential conflict if the unilateral adoption of environmental regulations were to create obstacles to international trade agreements. They also specified that control requirements be normalized for all producers of the same chemical.

The 1987 guidelines were in negotiation for five years, and during that entire time the developing countries consistently invoked the principle of prior informed consent (PIC), while the industrialized nations resisted its inclusion in the formal guidelines. The working group of legal experts continued its efforts following adoption, and the principle of prior informed consent was included in the amended London Guidelines, the first time it had been accepted in an agreed text negotiated by governments. It would later appear as part of legally binding treaties, together with the contents of Principle 21 of the Stockholm Declaration, in both the Basel Convention and the Biodiversity Convention.

The complementary nature of the London Guidelines may be seen in the following review of relevant work of other international organizations.

In 1976 the International Labour Organisation (ILO) included chemical safety as an important component of its program to improve working conditions and the work environment, providing conventions and recommendations, codes of practice, technical information, and technical cooperation projects covering occupational health and the use of such substances as white lead and benzene. It also included a version of the prompt notification requirement of the Provisional Notification Scheme.

The FAO deals with a wide range of potentially harmful chemicals, including pesticides, fertilizers, veterinary products, food additives, and food processing aids. In November 1985 the FAO adopted the International Code of Conduct on the Distribution and Use of Pesticides. In addition to including a prompt notification requirement, the code in 1989 was amended making banned or severely restricted pesticides subject to the doctrine of prior informed consent. After the adoption of the amended London Guidelines, the UNEP and the FAO developed a joint program to provide operational assistance, training, and technical advice to governments for implementation of the PIC procedure.

The World Health Organization (WHO) established the International Program on Chemical Safety (IPCS), one of a number of programs related to chemicals in the human environment. These include control of chemicals in the workplace, control of environmental pollution, use of chemicals in disease vector control, and control of chemicals in food.

The GATT is connected to the London Guidelines in two areas. The first is an Agreement on Technical Barriers to Trade, whose aim is to prevent unnecessary obstacles to trade that might result when a government or other entity adopts standards for protection of its own safety, health, or environment; the second is the working group established in 1989 to examine cases in which a contracting party, having banned the sale of a product domestically on health or safety grounds, seeks to sell it abroad.[3]

The Organization for Economic Cooperation and Development (OECD) began work on the control of chemicals in 1971, establishing what would be known as the OECD Chemicals Group. In 1973 it established the Committee on Consumer Policy, which operates an informal notification procedure on safety measures for hazardous products. In

1978 the OECD created a management committee for a special program on the control of chemicals; in 1984 it adopted a recommendation on an information exchange on export of banned or severely restricted chemicals.

Work is being done by other UN groups and other intergovernmental organizations. In 1982, the UN General Assembly called for the preparation and regular update of a consolidated list of products whose consumption or sale has been banned, withdrawn, severely restricted, or not approved by governments. At its 1984 meeting it decided that an updated consolidated list should be issued annually, in a form to permit computer access. (See appendix B to this chapter.)

In 1967 the European Economic Community (EEC) adopted a directive to harmonize the laws, regulations, and administrative provisions within the EEC on classification, packaging, and labeling of dangerous substances; in 1988 the European Community Council adopted a regulation dealing with the export and import of certain hazardous chemicals.

The foregoing review shows clearly that governments are deeply concerned about the international trade in hazardous chemicals, even though they have yet to negotiate a binding treaty that will control it. The issue has been ripe for consideration since 1989; in spite of numerous appeals and probably because of unwillingness on the part of the chemical producers, only in 1997 was action taken by the UNEP Governing Council to begin negotiating such a treaty.

The review also shows that governments will generally agree to negotiate the guidelines and principles referred to as "soft laws" when a potential source of conflict is recognized that cannot be solved unilaterally and when scientific, technical, and economic data show that the problem requires action, as long as the solution serves the interests of the governments, even though they are not ready to negotiate a binding treaty that would address the problem. Historically these soft laws follow a common pattern: They follow Principle 21 of the Stockholm Declaration, and they explicitly address issues of early notification; environmental impact assessments and their content, liability, and compensation; technology transfer; capacity building; disclosure to the public of potential hazards; and prior informed consent. However, still to be included in the legally binding agreements that have been developed are issues of transboundary environmental impact assessment and the public's right to information.

Notes

1. These are described in detail in Rummel-Bulska 1991a, 1992; UNEP 1978, 1980, 1982, 1984, 1985, 1987, 1989.

2. Agreements making specific requirements for various types of environmental impact assessment include Mediterranean Convention's Protocol for the Prevention of Pollution of the Mediterranean Sea by Dumping from Ships and Aircraft, February 1976, Article 7; Kuwait Regional Convention for Cooperation on the Protection of the Marine Environment from Pollution, April 1978, Article xi; (Abidjan) Convention for Cooperation in the Protection and Development of the Marine and Coastal Environment of the West and Central African Region, March 1981; (Lima) Convention for the Protection of the Marine Environment and Coastal Area of the South-East Pacific, November 1981; (Jeddah) Regional Convention for the Conservation of the Red Sea and of the Gulf of Aden Environment, February 1982, Article xi; United Nations Convention on the Law of the Sea, December 1982, Article 206; (Cartagena) Convention for the Protection and Development of the Marine Environment of the Wider Caribbean Region, March 1983, Article 12; (Nairobi) Convention for the Protection, Management and Development of the Marine and Coastal Environment of the Eastern African Region, June 1985, Article 13; ASEAN Agreement on the Conservation of Nature and Natural Resources, July 1985, Article 20, Paragraph 3(a); (Noumea) Convention for the Protection of the Natural Resources and Environment of the South Pacific Region, November 1986, Article 16; Convention on the Regulation of Antarctic Mineral Resource Activities, June 1988, Articles 4, 26(4), and 44(2).

3. In 1989 the UNEP Governing Council, at the urging of the authors, asked for a treaty covering such cases.

Appendix A: Conclusions of the Study of Legal Experts Concerning the Environment Related to Offshore Mining and Drilling within the Limits of National Jurisdiction

1. States should, either individually or jointly, by all appropriate means, take preventive measures against, limit, and in so far as possible reduce pollution and other adverse effects on the environment resulting from offshore exploration for and exploitation of hydrocarbons and other minerals, and related activities, within the limits of national jurisdiction. To this end, states should, in particular, adopt legislative and regulatory measures and provide appropriate machinery.

2. The important features of operations . . . should be made subject to a prior written authorization from the competent authority of the state.

3. The granting of an authorization should be preceded by an assessment of the effects of the proposed operation on the environment.

4. The assessment referred to in the conclusion above should cover the effects of operations on the environment, wherever such effects may occur. It should when deemed appropriate contain the following:

a. a description of the geographical boundaries of the area within which the operations are to be carried out

b. a description of the initial ecological state of the area

c. an indication of the nature, aims, and scope of the proposed operations

d. a description of the method, installations, and other means to be used

e. a description of the foreseeable direct and indirect long- and short-term effects of the operations on the environment, including fauna, flora, and the ecological balance

f. a statement setting out the measures proposed to reduce to the minimum the risk of damage to the environment from carrying out the operations and, in addition, possible alternatives to such measures

g. an indication of the measures to be taken for the protection of the environment from pollution and other adverse effects during and at the end of the proposed operations

h. a brief summary of the assessment that may be easily understood by a layman

5. Whenever a state has reason to believe that operations could have significant adverse effects on the environment of other states or of areas beyond the limits of national jurisdiction, it should provide such other states, as well as competent international organizations, with timely information that would enable them, where necessary, to take appropriate measures.

6. States should, by appropriate measures, provide for the determination of a person or persons, physical or juridical, to be liable for damage that may result from operations. The operator should be liable unless otherwise provided. Where more than one person is liable, their liability should be joint and several.

7. A state should assure to any person who has suffered damage as a result of operations an enforceable right to prompt and adequate compensation from the person or persons referred to in the previous paragraph, bearing in mind, inter alia, the degree to which such person may have contributed to the damage.

Appendix B: Internet Sources of Information on Banned or Restricted Chemicals

Title	Address	Document Size
PIC home page	http://irptc.unep.ch/pic/	3.3K
Pollutant Release and Transfer Registers (PRTR)	http://irptc.unep.ch/prtr/	2.3K
Implementation of the existing voluntary PIC procedure	http://irptc.unep.ch/pic/volpic/h2.html	3.4K
Background: PRTR histories	http://irptc.unep.ch/prtr/bakgd02.html	2.6K
Persistent Organic Pollutants (POPs) introduction	http://irptc.unep.ch/pops/intro01.html	6.2K
POPs: Socioeconomic consideration for global action, IFCS, June 1996	http://irptc.unep.ch/pops/indxhtms/manpops2.html	15.8K
POPs bibliography	http://irptc.unep.ch/pops/bibli0.1html	15.8K
Background on PRTRs	http://irptc.unep.ch/prtr/bakgd0.html	3.2K
International PRTR ativities	http://irptc.unep.ch/prtr/intl01.html	3.7K
Development of an international legally binding instrument for the application of the PIC procedure	http://irptc.unep.ch/pic/h2.html	4.1K
Further measures to reduce the risks from a limited number of hazardous chemicals	http://irptc.unep.ch/pic/furmer/h2.html	2.3K
Final report of the IFCS ad hoc working group on POPs, July 1996	http://irptc.unep.ch/pops/indxhtms/manwgrp.html	38.9K
Final Report of the IFCS ad hoc working group on POPs	http://irptc.unep.ch/pops/manila/manwgrp.html	38.3K

3

Negotiating Binding Regional Regimes: The Regional Seas Program

Oceans cover 71 percent of the world's surface and, through their interactions with the atmosphere and biosphere, play a major role in making life on Earth possible. They provide a habitat for a vast array of plants and animals and are a major source of human food, energy, and mineral resources. Keeping their waters productive is a vital part of any sustainable development strategy.

The vastness of the oceans gave rise to the myth that they had an infinite capacity to absorb and dilute whatever was put into them, leading to their being treated as a dump for all mankind's waste. Pollution has become a serious problem, especially in coastal areas and enclosed or semienclosed seas. Seven out of ten people around the globe live within 80 kilometers of the coast; almost half the cities with more than a million inhabitants lie at the mouths of tidal rivers. Coastal zones provide 90 percent of the world's fishing catch; in many countries fish is the major source of animal protein, accounting for 55 percent in Asia, for example. Mudflats, coral reefs, shallows, estuaries, caves, mangrove swamps, and beaches are the home of most marine life: A large majority of the 20,000 known varieties of fish, 30,000 species of mollusks, and almost all the crustaceans are found in the coastal zones, where, apart from ourselves, many birds and animals rely on the sea's harvest.

It is precisely there, where the sea is richest, that we humans put the most pressure on the marine environment, using it for our settlements, our food store, our playground, and our rubbish dump. Less than 10 percent of all the material entering these waters reaches the open ocean; the rest remains in the coastal sediment. Sludge from domestic sewers has almost completely stifled once productive shellfish beds near several North

American cities; in Indonesia, dramatic declines in shrimp harvests have been recorded in areas cleared of mangrove swamps. Perhaps 6.5 million tons of litter finds its way into the sea annually. In the past, much of this disintegrated quickly, but resistant synthetics have replaced many natural, more degradable materials. Plastics, for example, persist for up to 50 years and, because they are usually buoyant, are widely distributed by ocean currents and the wind. Most beaches near population centers are littered with plastic waste from rivers, ships, outfalls, or illegal refuse operators or left behind by beach users.

Another important source of shoreline pollution is oil. It was estimated in 1985 that 223 million barrels—3.2 million tons—of oil entered the marine environment every year. Of this, municipal waste and runoff accounted for 8.1 million barrels and maritime vessels for about 10.1 million barrels. By 1989 the latter figure had fallen considerably, with oil pollution from sea traffic at about 4 million barrels, of which 20 percent came from tanker accidents. This improvement stemmed largely from the adoption of the International Convention for the Prevention of Pollution from Ships, adopted in 1973 and modified by the Protocol of 1978, which entered into force in 1983. The Convention, known as MARPOL 73/78, now applies to more than 85 percent of the world's merchant fleet.

Red tides caused by algae blooms are annual events in many parts of the world. Japan's Inland Sea is affected by some 200 red tides each year. Blooms of toxic species occurred in the North Sea with increasing frequency in the 1970s and 1980s; in 1988 a massive bloom occurred in the seas around southern Scandinavia, damaging marine life in some seas and affecting some fish farms along the Norwegian coast. Although unusual occurrences of algae blooms have been attributed to a combination of many factors, especially to disturbances in the marine ecological balance caused by climatic factors, considerable evidence suggests that their increased incidence is related to the nutrient enrichment of coastal waters and inland seas on a global scale.

The main risk to human health associated with marine pollution is from domestic sewage discharged into coastal waters. Eating contaminated seafood is firmly linked to serious illnesses, including viral hepatitis and cholera; epidemiological studies have proven that people who swim in sewage-polluted waters have an above normal incidence of gastric disorders; ear, respiratory, and skin infections may also result. Many countries, including France, Greece, Italy, and the United States, have had to close beaches temporarily because their water quality was unacceptable for swimming.

A challenge facing the world community is how to strike a balance between development and the preservation of the coastal environment. Most waste, once introduced into the sea, cannot be removed. Its future course is determined by its chemical composition and decomposition rate and by the currents that carry it. Nondegradable waste may travel long distances.

What, then, should be done with the waste from our factories, homes, and ships? It must go somewhere. Yet we must consider our neighbors as well as ourselves. Scores of people in the Japanese village of Minamata died when they ate fish contaminated by waste from a nearby industrial plant; traces of the pesticide DDT have been found in Antarctic penguins and seals, far from their source. Coastal ecosystems intermesh with terrestrial and open ocean systems to form a complex whole. The tides and currents that make coastal waters so productive also render them exceptionally vulnerable to pollution: Damage one part of the system and a chain reaction begins. Deforestation in watersheds hundreds of kilometers inland can result in silt-choked harbors and dying coral reefs.

Clearly, environmental problems rarely affect one nation alone, especially when they involve coastal areas and the marine environment. Because regional fishing grounds are normally shared by several nations, each country's pollution can degrade the shared environment and damage the resources of all. Oil spills, land-based pollution and the pressure of human settlements on animal habitats are often specific to a region; protecting the coastal environment and the sea's resources therefore depends on regional cooperation.

Historically, international marine agreements regulated navigation and fishing; it has only recently been recognized that the world's oceans should be regulated and protected as a natural resource. This important change from a user-oriented to a resource-oriented approach has come about only in the last two decades. Most legal regimes adopted since 1971 have included the protection, conservation, and management of the marine and coastal environment and their resources.

In the late 1960s and early 1970s, scientific findings were widely publicized that the Mediterranean was a dying sea, that the Caribbean Sea and the Arab/Persian Gulf were heavily polluted, and that the Pacific fishing grounds had been overexploited. Alarmed, governments in these areas cooperated to find durable solutions. With the UNEP acting as catalyst and coordinator, the Regional Seas Programme was launched in the mid-1970s, its basic strategy to deal with the causes as well as the effects of coastal environmental damage.

The Mediterranean

A miniature ocean bordered by 120 cities with a population totaling at least 100 million, the virtually enclosed waters of the Mediterranean Sea have been the crossroads of European, Asian, and African civilizations for at least 4,000 years of recorded history, but by the early 1970s the Mediterranean was so heavily polluted that many feared it might die. Once a symbol of the seas' benefits to man, it became a symbol of man's destructive impact on the seas. Efforts to save it began with an assessment of its condition, carried out by a team of technicians from all the relevant UN organizations. Their prognosis was bleak.

The question then became, in the midst of wars, political antagonisms, and national feuds, to what extent would countries around the Mediterranean be willing to enter into an environmental agreement that would benefit them all? This was a time when all the Arab states were at war with Israel, Turkey and Greece were disputing ownership of Cyprus, Algeria and Morocco were at odds over the Sahara, and the Cold War was still shaping international relations. In spite of these difficulties, and in the face of the belief that the Mediterranean was beyond saving, the UNEP decided to go forward.[1] Spain offered to host meetings to negotiate regional cooperation in an effort to save the Mediterranean and to the astonishment of many, almost all of the basin states not only attended the negotiating sessions, but also in 1975 succeeded in adopting a joint action that would slow and ultimately reverse the threat.

In *Saving the Mediterranean*, Peter Haas (1990) said,

Countries disagreed about the thrust of the program and about what would constitute appropriate supporting institutional arrangements. In Barcelona, Less Developed Countries (LDCs) argued for a program which would enhance their marine science capabilities. They wanted a regional operational center that would perform both a switchboard function for the transmission of information and coordination and other executing functions, as well as actually providing technical assistance and transferring pollution monitoring equipment to the LDCs. They also supported the development of comprehensive regional arrangements for pollution control which would be legally binding on the participants. Spain, Italy and France, already possessing effective marine laboratories, and represented with very powerful and highly qualified delegations, wanted only minimal, flexible, and mostly subregional cooperation schemes, and preferred a weak organization that would only facilitate information exchange. They thought that further responsibilities should be purely voluntary and bilateral.

The governments represented responded to the public outcry particularly in the developed countries of the north Mediterranean. Yet, because of these disagreements, the meeting approved stronger monitoring and assessment proposals than the developed countries wished, but the supporting administrative arrangements remained unspecific.

Haas also points out that a conflict between the Arabs and Israelis was avoided by their agreement that the Arab states would not challenge Israel and the Israeli delegates would maintain a very low profile (the PLO was not invited to the meeting). The United States attended as an observer, but only until it was convinced that the Mediterranean Plan would not affect its Sixth Fleet deployment.

A disagreement that for a time threatened the harmony of the negotiations arose around the question of where to locate the coordination unit of the plan. Spain and Greece lobbied intensely for this privilege, offering very tempting facilities and financial support; at last, informal consultations and a straw ballot showed that Greece would win, whereupon Athens was formally chosen. For the most part the continuing official meetings have been marked by adherence to tacit rules of diplomatic behavior between the parties which have defused traditional sources of conflict. For example, the Turkish delegation to a meeting regularly submits a letter to the UNEP Secretariat protesting the Greek presence in Cyprus; the Secretariat lays the letter on a side table; with no action, there is no conflict. Similarly, when Syria ratified the Convention it recorded a qualification that its ratification did not constitute recognition of Israeli sovereignty. And delegates refrain from criticizing other countries' polluting habits, in an implied recognition that all are guilty. This is also due to the nature of the negotiations, which were intended rather to save the Mediterranean than to assign blame.

It must be admitted that some political considerations prevailed over efforts to control pollution. When the parties were assigning functions to various locations around the Mediterranean, Greek and Turkish delegates opposed locating any institutions in each other's countries. More seriously, delegates of most of the Mediterranean states, particularly the north Mediterranean, were clearly unwilling to admit an effective USSR presence. For this purpose they defined the Mediterranean as the area between the Straits of Gibraltar and the southern limits of the Dardanelles between the Mehmetick and Kumkale lighthouses in Turkey, thus eliminating the Black Sea and rivers flowing into the Mediterranean and blocking participation of the Black Sea states of Bulgaria and Romania. Other nonlittoral states with rivers emptying into the Mediterranean, such as Portugal, Switzerland, and the Sudan, were by this definition excluded from the agreement. This reduced the effectiveness of the Convention, because it cannot deal with some major sources of pollution, including the Gulf of Izmit and the Sea of Marmara.

For the UNEP, the most encouraging aspect of the Mediterranean negotiations was the political breakthrough, which did not mean, however, that results flowed automatically or easily. Political barriers remained, and it was impossible to get the forty-odd participating institutes to work together. It was necessary to establish seven networks, one or more of them under the aegis of a specialized UN agency, principally the FAO and WHO. The networks were carefully designed to avoid political conflict—in none of them was it necessary for Arab and Israeli institutes to work together—while ensuring that the flow of information and data would continue to show everyone in the region the seriousness of the situation.

These arrangements worked so successfully that the UNEP team was encouraged to proceed with a legally binding convention. A year of intensive negotiations, during which there was a general agreement that the parties needed to work together, led in 1976 to the adoption and signing of the Barcelona Convention for the Protection of the Mediterranean Sea against Pollution and two protocols, one to prevent pollution of the Mediterranean by dumping from ships and aircraft and the other to achieve cooperation in combating pollution by oil and other harmful substances. For the agreement to enter into force it was required that the participating states ratify the Convention and at least one of the two protocols, so that the countries were committing themselves to a general cooperation and to at least some cooperation in specific areas.

In 1979 a plan was launched that would integrate development plans for environmental protection in the Mediterranean, the so-called Blue Plan, which was part of the socioeconomic component of the Mediterranean Action Plan. In 1980 the region's states adopted a protocol identifying measures to control land-based pollution of the sea by municipal sewage, industrial waste, and agricultural chemicals; two years later they approved a further protocol to protect endangered species of fauna and flora and their habitats.

Yet environmental improvements were slow to come. By the tenth anniversary of the Barcelona Convention, accomplishments seemed meager; at a meeting in 1985 in Genoa, and in spite of delaying tactics of the industrialized countries led by France, ten priorities were set for the decade 1985–1995. Three of these, preceding the global and regional accords on ozone depletion and sulfur dioxide production, were the first time-constrained goals to be adopted by a group of nations: establishment of sewage treatment plants in all cities in the region having more than 100,000 inhabitants and appropriate outfalls/treatment plants for all towns with more than 10,000 inhabitants; identification and protection of at least 100 sites of common interest; and identification and protection of at least fifty new marine and coastal sites and reserves. Also subject to time deadlines were the establishment of reception facilities for contaminated ballast waters and other oily residues in Mediterranean ports; application of environmental impact assessments; risk reduction in the transport of dangerous and toxic substances through cooperative maritime safety measures; protection of endangered marine species such as the monk seal and the Mediterranean turtle; measures that would substantially reduce industrial pollution and solid-waste disposal in the sea; intensification of efforts to prevent and combat forest fires, soil loss, and desertification; and a substantial reduction in air pollution, with its adverse effects on the coastal and marine environments and its accompanying danger of acid rain.

In 1991, at the fifteenth anniversary meeting of the Barcelona Convention, former UNEP Executive Director Mostafa Tolba called attention to the basic issue of all environmental treaties:

> A very basic question we have to answer is how far has the Mediterranean benefited from the Barcelona Convention, its various protocols, and its action plan. We are all saying the Mediterranean would have been worse without them. We need the proof for this. We need to know exactly where we were and how far did we go. This is essential to identify the next concrete steps. My question is, how far are the contracting parties ready to support such an exercise financially and with human resources? Such an effort will be a huge multidisciplinary effort involving marine scientists, ecologists, economists, technologists, social scientists, and several others.

Accords Extended to Other Regional Seas

What happened in the Mediterranean had a significance beyond its shores. As the program evolved, the participants were able to enlarge their understanding of the environment's role in development and began to see evidence that both developed and developing nations were prepared to put aside their political differences as they cooperated to protect their shared environment. As the field of environmental diplomacy developed, several action plans, conventions, and protocols were adopted.[2] Now, through the Regional Seas Programme, some 130 countries, 16 UN agencies and more than 40 other international and regional organizations are working together to improve the marine environment and make use of its resources.

Although the formal requirements for achieving each action plan would seem to be enough to occupy its negotiators, in fact difficulties completely outside the agreements themselves often postpone, delay, or derail them. As an example of the knotty political problems that arise, let us consider the Kuwait Action Plan. The seven riparian Arab states and Iran, although agreeing that a convention was necessary to assure the ecological health of the body of water in question, could come to no agreement as to its name. The Arabs insisted it should be called "the Arab Gulf"; the Iranians were adamant that it would go under the UN term, "the Persian Gulf"; suggestions for compromises such as "Arab/Persian Gulf" or simply "the Gulf" were rejected by both sides. In order to get on with the negotiations it was finally agreed to call it "the body of water surrounded by Bahrain, Iran, Iraq, Kuwait, Oman, Qatar, Saudi Arabia, and United Arab Emirates," and, since the meeting was to be held in Kuwait, to name the adopted convention the Kuwait Action Plan Convention. The long, clumsy working title was adopted, the political impasse overcome.

Similarly, the negotiations for the Caribbean Action Plan were complicated by political sensitivities. European countries having territorial claims in the Caribbean attended the meetings, and from time to time the Caribbean states would exclude them and hold their own meetings, creating very embarrassing situations. But patience and the common desire to protect the region prevailed, and the Caribbean Action Plan was followed by the Cartagena Convention. Interestingly, once the Kuwait and Cartagena agreements were reached, Iran and the European countries agreed to pay the largest shares to their respective funds to finance implementation.

In West and Central Africa, the main obstacle was the location of the coordinating unit that would implement the convention and the action plan. Senegal, Côte d'Ivoire, and Nigeria were the main contenders for what they saw as a base of power, and after a good deal of wrestling with the issue they came to a compromise: a periodic sharing, beginning with the country where the treaty was signed, Côte d'Ivoire.

Most of the regional seas treaties provide for cooperation among states to deal with pollution emergencies, but none defines what constitutes an emergency. One protocol, the Protocol Concerning Cooperation in Combating Pollution of the Mediterranean Sea by Oil and Other Harmful Substances, provides that the Mediterranean states will cooperate ". . . in cases of grave and imminent danger to the marine environment, the coast or related interests of one or more of the Parties due to the presence of massive quantities of oil or other harmful substances resulting from accidental causes or an accumulation of small discharges which are polluting or threatening to pollute the sea . . ." (Article 1). However, there are no definitions of "grave," "imminent," or "massive."

Even the term "emergency" has yet to be defined in most treaties, although the "related interests" referred to in the quotation above include the health of the coastal population; maritime, coastal, port, or estuarine activities; fishing, management, and conservation of living natural resources; historical and tourist appeal of the area, including water sports and recreation; and the cultural value of the area. Similarly, the definition of a marine pollution incident remains ambiguous, with the Caribbean Protocol defining only the oil spill incident and the Eastern African Protocol containing a broader definition: "A discharge or spillage of oil or other harmful substances into the marine environment, or a significant threat of such a discharge or spillage, however caused, of a magnitude that requires emergency action or other immediate response for the purpose of minimizing its effects or eliminating the threat" (Article 1).

In spite of these ambiguities, in an implicit move away from the doctrine of absolute state sovereignty, all the conventions concluded under the UNEP's Regional Seas Programme contain an article dedicated to the obligation to inform endangered states and the competent international bodies when an emergency occurs. For example, Article 5 of the Eastern Africa Protocol requires its parties to establish procedures to report marine pollution incidents as rapidly as possible, and an annex to the Protocol provides guidelines for doing so.

However, in spite of these encouraging signs, progress has been slow, especially in developing regions, where most countries lack the capacity to assess their marine and coastal environments or to manage their resources rationally. Weak institutional structures hamper their participation in international efforts; the lack of resources makes it difficult to respond rapidly and effectively. Without the necessary material and training resources, regional agreements will be of limited use.

What Have We Learned?

The negotiation of ten or more regional seas action plans and more than 25 legally binding regional agreements has yielded useful information for future negotiations. To begin with, when science speaks with authority, governments listen. Scientific reports that identified the cause of the Mediterranean's ills spurred the countries surrounding it to action; subsequent action plans and agreements have been preceded by scientific assessment of the regional sea in question. Governments have also shown themselves to be willing to put aside political differences and address a common threat, but such negotiations succeed only when they share certain features: there must be strong leadership by at least one of the parties; the sponsoring UN organization must take an active, objective role in the meetings; and the negotiating delegates must be made up of government representatives whose strong personalities lead them to make imaginative, effective decisions.

Dealing with shared environmental problems has inevitably led to an erosion of the old doctrine of absolute sovereignty, as governments have become willing both to give and to accept instructions as to how to modify pollution-causing activities. But the resulting treaties are not, in themselves, enough. Implementation is the key. Any successful treaty must provide for enough financial and technical resources to countries that need help in enforcing the terms of the treaty inside their own borders.

Last, the importance of public awareness and sensitivity cannot be overemphasized. In the case of the Mediterranean Sea, the public outcry that followed media coverage of Jacques Cousteau's warnings along with media coverage of further scientific findings prompted the region's governments to act in concert to avert the environmental calamity. The same pattern saved the waters in the region covered by the Kuwait Convention.

Notes

1. Universal attendance was achieved by a team consisting of Mostafa Tolba; Peter Thacher, the American head of the UNEP's Geneva office; Stephan Keckec, a Yugoslav and the head of the UNEP's Mediterranean Program; Patricia Bliss, an American lawyer; and Mohamed Tanji, a Moroccan marine biologist.

2. A summary of these is as follows: (a) Kuwait Action Plan region: Kuwait Regional Convention for Cooperation on the Protection of the Marine Environment from Pollution (Kuwait, 1978) and Protocol Concerning Regional Cooperation in Combating Pollution by Oil and Other Harmful Substances in Case of Emergency (Kuwait, 1978); (b) West and Central Africa: Convention for Cooperation in the Protection and Development of the Marine and Coastal Environment of the West and Central African Region (Abidjan, 1981) and Protocol Concerning Cooperation in Combating Pollution in Case of Emergency (Abidjan, 1981); (c) South-East Pacific: Convention for the Protection of the Marine Environment and Coastal Area of the South-East Pacific (Lima, 1981), Agreement on Regional Cooperation in Combating Pollution of the South-East Pacific by Hydrocarbons or Other Harmful Substances in Cases of Emergency (Lima, 1981), and its Supplementary Protocol (Quito, 1983), Protocol for the Protection of the South-East Pacific against Pollution from Land-Based Sources (Quito, 1983), Protocol for the Conservation and Management of Protected Marine and Coastal Areas of the South-East Pacific (Paipa, 1989), and Protocol for the Protection of the South-

East Pacific against Radioactive Contamination (Paipa, 1989); (d) Red Sea and Gulf of Aden: Regional Convention for the Conservation of the Red Sea and Gulf of Aden Environment (Jeddah, 1982), and Protocol Concerning Regional Cooperation in Combating Pollution by Oil and Other Harmful Substances in Cases of Emergency (Jeddah, 1982); (e) The Caribbean: Convention for the Protection and Development of the Marine Environment of the Wider Caribbean Region (Cartagena, 1983), Protocol Concerning Cooperation in Combating Oil Spills in the Wider Caribbean Region (Cartagena, 1983), and Protocol Concerning Specially Protected Areas of Wildlife to the Convention for the Protection and Development of the Marine Environment of the Wider Caribbean Region (Kingston, 1990); (f) East Africa: Convention for the Protection, Management and Development of the Marine and Coastal Environment of the Eastern African Region (Nairobi, 1985), Protocol Concerning Protected Areas and Wild Fauna and Flora in the Eastern African Region (Nairobi, 1985), and Protocol Concerning Cooperation in Combating Marine Pollution in Cases of Emergency in the Eastern African Region (Nairobi, 1985); (g) South Pacific Region: Convention for the Protection of the Natural Resources and Environment of the South Pacific Region (Noumea, 1986), and Protocol Concerning Cooperation in Combating Pollution Emergencies in the South Pacific Region (Noumea, 1986). Other action plans, conventions, and protocols are at various stages of development.

4

Negotiating Binding Regional Regimes: Shared Freshwater Resources

The difficulties encountered in negotiating the regional seas agreements are matched and sometimes exceeded by those of reaching agreement on management of shared freshwater resources: rivers, lakes, and aquifers, including their living aquatic resources. Fresh water is one of the basic requirements of life, and in theory there is enough of it to meet the needs of all, but in practice it is often a scarce resource.

Some human settlements have grown up where there is ample fresh water, others where there is more water than they need or want, but not in the right place or at the right time, still others where there is barely enough water, compelling them to exist in a perpetual state of drought. Some of this imbalance may be traced to changing global weather patterns, but some is due to the modification of water circulation and quality. Development has led to overexploitation of freshwater resources and to channelization and other regulation of river courses, and changing land use patterns have led to changes in vegetation and soil cover that have affected the hydrological cycle.

The near destruction of Lake Erie by industrial pollution, the poisoning of more than 20,000 lakes in Scandinavia and Canada by acid rain, the siltation and eutrophication of thousands of lakes in developing countries by imprudent development and sanitation practices—all these merely hint at the extent of the problem. Traditional cost-benefit judgments in these cases may be inappropriate: for example, when lakes near industrial parks are sacrificed in the name of economic development or nondegradable toxic chemicals are discharged to avoid the expense of pollution control.

In 1950 it was estimated that the world used about 1,360 cubic kilometers of fresh water. By 1990 that figure had increased to 4,130, and it is expected to reach about 5,190 cubic kilometers by the year 2000, nearly a fourfold increase. Of this total, 69 percent goes for agriculture, 23 percent for industry, and 8 percent for domestic uses.

Many of the world's freshwater resources are shared by two or more states. Using the national boundaries that preceded the breakup of the Soviet Union and other Eastern European countries, it was found that there were 214 multinational river basins: 155 shared between two countries, 36 among three, 23 among four to twelve. More than a third of the world's population lives in international river basins, and 75 percent or more of the area of some 50 countries lies within them.

The joint use of international watercourses has always depended on cooperation among the countries along their banks, regulated in some cases by international treaties and organizations. Historically, these treaties dealt with the allocation of water shares, regulation of navigation and fishing, and construction of dams and other public works. As early as 1792 the Provisional Executive Council of the French Republic stated, "Streams that flow through the territory of several states do not belong exclusively to these states with respect to the particular portion traversing this territory, but represent the common and inalienable property of all riparian states." This principle was echoed by the 1814 Treaty of Paris. Other recent agreements, especially beginning in the 1970s, have focused on the cleanup of shared waters. The Great Lakes Water Quality Agreements of 1972 and 1978 focused respectively on traditional pollution sources such as municipal sewers that were causing severe eutrophication and on toxic pollutants. Since 1980 countries along the Rhine have undertaken a joint program for the rehabilitation of its waters and the management of its aquifer.

In other regions the shortage of freshwater resources may result in conflicts among the nations that share them. A report published by Washington's Center for Strategic and International Studies states, "By the year 2000, water—not oil—will be the dominant resource issue of the Middle East. If present consumption patterns continue, emerging water shortages, combined with a deterioration in water quality, will lead to more desperate competition and conflict." In the sub-Sahara, water management and planning is complicated by recent climatic changes and the expected global warming. Recent studies show that rainfall has been continuously in decline for the past twenty years, especially in Senegal, Mali, Burkina Faso, and Niger, and analyses of monthly rainfall show the most marked decline in August, a critical month for rain-fed agriculture. Areas of the Sahel that depend on irrigation are also experiencing drought: The average annual flow of all the rivers reaching the Sahel declined by some 25 percent between 1968 and 1983, a deficit equivalent to three times the capacity of the Manatali Dam on the Upper Senegal River; between 1965 and 1983 in places like Maradi (Niger) the annual total flow never once reached the long-term average for 1931 to 1960.

International tensions during the last decades have centered on the development of shared rivers. The United States and Mexico have clashed over the Colorado River; Iraq, Syria, and Turkey over the Euphrates; India and Pakistan over the Indus; Israel and Jordan over the Jordan; Brazil and Argentina over the La Plata. The scope for conflict will rise with the growth of populations, and opportunities must be created for mutual cooperation, dialogue, and confidence building among rivals who must share this vital resource.

The picture is not, however, unrelievedly dark. The experiences of UNEP representatives in advancing the Regional Seas Programme showed that international concern for environmental issues can triumph over political and economic conflicts. The organization was thus encouraged to enter the more complex and problematic area of the management of shared freshwater resources (including their allocation, use, and conservation, as well as their protection against pollution), beginning in 1985 with the Zambezi Action Plan.

The Zambezi Action Plan

Eight countries affect the waters of the Zambezi River. Arising near the extreme northern border of Zambia and Angola, this great river becomes the border between Zambia and Namibia, Zambia and Botswana, and Zambia and Zimbabwe. Fed by tributaries from Tanzania and Malawi, it then flows through Mozambique to the Indian Ocean. In its course, which leads it more than halfway across southern Africa, the Zambezi collects waters that more than fill the needs of the riparian states; nevertheless negotiation of the Zambezi Action Plan and its accompanying agreement met a series of nearly insurmountable difficulties.

At the time the plan was initiated, the southern African states had established the Southern African Development Co-ordination Conference (SADCC), now called the Southern African Development Community (SADC). Another difficulty was that the political situation in South Africa had eroded confidence between some of the governments involved, who suspected others of intending to pass some of the Zambezi's water to Southern Africa. The Conference Secretariat was at best lukewarm to the UNEP's proposal, and it became evident that to be effective, the plan would have to have the support not only of the highest political levels of government, but also of the donor community, because the riparian states were all poor.

Visits were arranged in which UNEP Executive Director Mostafa Tolba met with the heads of state in three of the major countries involved: President Museri of Botswana, President Kaunda of Zambia, and President Banana of Zimbabwe. The concept of regional management of the Zambezi basin was endorsed by all three leaders, who gave the UNEP authorization to proceed. With this important endorsement, the UNEP[1] approached the donor countries.

Although they had at that time little experience working with the kind of men who governed some of the nations involved in the project, three Nordic states (Finland, Norway, and Sweden) and Canada expressed interest. They provided high-level experts who helped the UNEP senior staff work with the national experts of the riparian states, and by the time the action plan had been drawn up, the Nordic states had gone as far as to commit themselves to furnish the $10 million it was estimated to cost.

Preparation of the action plan began with a diagnostic study of the river basin, including information on river flows, national water needs, development plans, and conservation. Although this took a long time, as a straightforward collection of factual information it was clear and powerfully convincing to the riparian states involved, who agreed that cooperation was necessary.

Negotiations were protracted and difficult, mostly for reasons having to do with national political interests (which remained high on the local experts' agendas) and the various nations' attitudes toward events in southern Africa. It took a little more than two years, from April 1985 to May 1987, to reach an agreement. Technical meetings were held in Nairobi, Lusaka, Gaberone, and Harare in which each expert presented technical and scientific evidence to support his or her point of view.

Even more difficult were the negotiations held at the ministerial level, in Harare in May 1987, that would adopt the action plan and its implementing legal agreements. The meeting was led by Victoria Chitepo, then Zimbabwe's Minister of Natural Resources and Tourism. Chitepo, Tolba, and Iwona Rummel-Bulska worked closely together to guide the consultations. Of the eight riparian states, only six attended; governments lacked confidence in each other, and it was difficult even to establish the drafting and credentials committees. It took more than a full day of the three days allotted to the meeting to agree that five countries would sit on the drafting committee and all six on the credentials committee. The composition of the bureau of the ministerial conference was sensitive, and Zimbabwe, Zambia, and Botswana were chosen to ensure balanced views. During the ministerial meetings the value of informal consultations once again proved itself. The potential for political differences to erupt was always present, and the UNEP officials were at pains to remain objective and not to appear to favor any party over another. This is a difficult, sometimes frustrating stance to maintain, but a necessary one for any international organization that acts as broker for international environmental negotiations.

It was evident during these meetings that since the late 1970s, when the Shared Natural Resources Guidelines were being negotiated, attitudes toward national sovereignty had changed. Instead of refusing to consider international control of shared waters, as Brazil had earlier done regarding the La Plata River, the riparian states of the Zambezi saw clearly that none could manage its part of the water alone and that all must cooperate. A few paragraphs of the approved action plan illustrate this new attitude:

> In view of the present utilization of the river system it is possible and highly desirable to deal with the water resources and environmental management prob-

lems of the river system in a coordinated manner to avoid possible future conflicts.

The diagnostic study identified, inter alia, the following problems relating to the environmentally-sound management of the river basin which should be dealt with through selected activities as part of the Zambezi Action Plan.

(a) Inadequate co-ordination both at national and at river basin level.

(b) Inadequate information on environmental impacts of water resources and related development projects, e.g., hydropower, irrigation, etc.

The objective of the ZACPLAN [Zambezi Action Plan] is to overcome the problems listed above and thus to promote the development and implementation of environmentally-sound water resources management in the whole river system. The activities in the river basin should include solid and reliable environmental assessment . . .; the information falling into this category should relate to:

(a) Socio-economic development that may adversely affect the environment, including the identification of favorable opportunities for river basin development in general.

(b) The identification of human activities that could be affected by environmental degradation.

The gradual development and operation of a basin-wide unified monitoring system for water and water-related environment.

Development of a river basin planning process.

Other Shared Water Resources: Lake Chad; the River Nile

The success of the Zambezi basin negotiations led the governments surrounding Lake Chad (Chad, Niger, Nigeria, and Cameroon) to work on the development of a similar plan. The process began in 1988, again with a diagnostic study. At this time the lake had shrunk on one shore by twenty-five kilometers, and on another by forty kilometers, but the causes were controversial. Unfortunately the objectivity of the study was called into question and it had to be completely redone; this and other factors caused the initial enthusiasm for the plan to fade. Even though in the end the ministers agreed to the plan, at this writing it has not been signed by the heads of all the riparian states.

While these efforts were under way, the UNEP stepped into the much more complicated situation of the Nile basin. The early phases of the negotiations were marred by hypersensitivity on the part of some delegates, and it took two full years of concerted effort by both water experts and environmental law specialists to get them to admit that collective management of the Nile basin for the benefit of all was the best option. Ultimately, they agreed to develop a diagnostic study and an action plan. The process is continuing, although slowly, at this writing.

The same slow progress is reported in achieving a joint action plan for the Río Arauca, shared by Colombia and Venezuela. Obviously, management of shared freshwater resources is a more difficult area of negotiation than management of regional seas, although it is encouraging that progress has been made toward agreeing to limit territorial sovereignty for the common good.

Note

1. The proposed agreement was the concern mainly of Tolba, Rummel-Bulska, and the UNEP's senior water specialist, the late Laszlo David of Hungary.

5

The Story of the Ozone Layer

During the past three decades governments, international agencies, and NGOs have developed increasingly effective approaches to transboundary environmental problems. From the traditional, insular stance by which a nation rejected interference in its policies, in the environmental field governments have realized that some problems must be solved through cooperation.

The earlier conventions and protocols, which responded to crises such as acid rain and dumping of hazardous wastes, were achieved through procedures that would prove valuable when it was realized that serious threats had existed undetected since as early as the 1970s. These included depletion of the ozone layer, global warming, climate change, and the loss of biodiversity. A combination of advancing scientific methods and knowledge and heightened awareness of the fragility of the environment uncovered all these problems, and the experience gained in earlier negotiations made it possible to move relatively swiftly to agreements that it was hoped would solve them.

Negotiations for the Vienna Convention and its Montreal Protocol, which addressed the problem of ozone depletion, set new standards in international relations and are therefore described at length. They extended from the first announcement of the existence of the problem in 1974 (see Rowland and Molina 1975) through the signing of the Vienna Convention in 1985 and the Montreal Protocol two years later to subsequent amendments of the Protocol in 1990 and 1992. Despite the agreements reached, there is still debate on the need to regulate ozone-depleting substances: Some states oppose regulation and cite scientific uncertainty as their justification, and the timetable for phasing out the suspect substances will be contentious for some time in the future.

Although most readers are aware of the significance of the ozone layer, it would be useful at this point to provide a brief description of its function. The atmosphere has evolved over geological time into a mixture of roughly 80 percent nitrogen and 20 percent oxygen, with many trace gases in concentrations from a fraction of 1 percent down to a few parts per trillion. These trace gases are much more important than their concentrations suggest. Ozone, in particular, though found in a concentration of only about ten parts per million by volume, acts as a natural filter, absorbing and blocking the sun's shortwave ultraviolet radiation (UV-B). Ozone is produced and destroyed at a wide range of altitudes in the atmosphere. Near the ground, where it is formed by the reaction of automobile and industrial emissions with sunlight, it is a harmful pollutant, but between twenty-five and forty kilometers above the earth, in the stratosphere, ozone emerges naturally, existing in equilibrium as it is formed from molecular oxygen and destroyed by UV-B radiation.

In the late 1960s there was concern over the then-new supersonic aircraft, whose stratospheric optimum cruising altitude and high engine temperatures allowed them to convert atmospheric nitrogen and oxygen into nitrogen oxides, which then acted to destroy ozone. It had been realized that a number of reactive chemicals (such as oxides of hydrogen, nitrogen, chlorine, and bromine) can act as catalysts, destroying ozone molecules sometimes by the thousands before being removed from the stratosphere through interaction with other free radicals. Before concern about the threat could be mobilized, however, plans to build fleets of commercial supersonic aircraft were abandoned, not from concern for the environment, but for cost-benefit reasons.

In 1975 Professors Sherwood Rowland and Mario Molina of the University of California pointed to the chlorofluorocarbons (CFCs) widely used in homes and industry (as propellants in spray cans, in cooling systems, in foam blowing, and as solvents in the electronics industry, for example) as endangering the ozone layer. Both chlorine and fluorine are highly reactive, making them particularly useful when combined for industrial purposes but also potentially destructive. As these substances rise through the atmosphere into the stratosphere, the researchers said, they cause a catalytic reaction which destroys ozone. Since it was already known that ozone blocks UV-B from entering the atmosphere and suspected that UV-B can cause cancer and cataracts in humans and adversely affect plant fertility, this report acted as a call for action. The UNEP Governing Council responded in 1976 by calling for a meeting of the appropriate international governmental and nongovernmental organizations to review all aspects of the ozone layer. However, it would take ten years' work to achieve a framework convention of general principles that would result in cooperative action.

The initial meeting, held in Washington in 1977, drew up a world plan of action containing research recommendations for the study of the natural ozone layer and changes in it caused by man's activities, as well as these changes' effects on man, the biosphere, and the climate. The UNEP and the World Meteorological Organization (WMO) established a Coordination Committee on the Ozone Layer made up of representatives of specialized agencies, national, intergovernmental and nongovernmental organizations, and scientific institutions. This committee was to produce semiannual assessments of ozone layer depletion and its impact.

As knowledge of ozone chemistry advanced, the Ozone Committee revised its estimate of ozone depletion. For example, in October 1981, the estimate was that there would be an eventual depletion of 5–10 percent, if emissions of chlorine compounds continued at then-current levels; this was a lower figure than previous estimates but was no cause for complacency. At that time there was some doubt that CFCs were a cause of ozone depletion, but the Canadian government was sufficiently concerned about these substances that it had already ordered the progressive elimination of the most common CFCs (labeled CFC 11 and CFC 12) and had prohibited their use in nonessential spray cans. In 1978 the United States had banned their use as aerosol propellants, which prompted some large chemical companies to begin a search for safe substitutes.

In 1981, accumulating scientific information led the UNEP Governing Council to establish an ad hoc working group of legal and technical experts to work out a framework convention on the protection of the ozone layer; the group met four times between January 1982 and March 1985. At the first meeting it was recognized that although the threat was still distant and the issue shrouded in scientific uncertainties, there was an urgent need for an international effort to deal with it. Excerpts from the opening statement by Mostafa Tolba provide an overview of the situation (UNEP 1983, p. 231):

We are meeting to lay the groundwork for a convention that some argue may not be needed, for we cannot say with certainty that the ozone layer is being depleted. However, all the most reliable scientific evidence points to the fact that the earth's protective ozone layer has been, is being, and, more importantly, will continue to be depleted by chlorofluorocarbons and other chemicals unless the international community takes preventive action. The urgent task of this meeting is to create a framework that will make that action effective.

Less than a decade ago, the problem of ozone depletion caused by chlorine atoms penetrating the stratosphere was hardly known. Now, following substantial research and monitoring, we are aware of a potential pollution problem of a scale and consequence never before faced.

The costs of continued chlorofluorocarbon pollution are not primarily ours to bear. If scientific observations over the next few years turn the theory of ozone depletion into unchallengeable fact, then the hazard of increased ultra-violet light exposure due to ozone depletion is a legacy we will pass on to future generations. Not only man but most other living things are susceptible to UV-B exposure. Nor is the problem one of limited area. Atmospheric transport ensures that the risk is distributed to all parts of the globe.

If ozone depletion does reach levels where it becomes measurable and at the same time chlorofluorocarbon emissions remain at their present levels then, because of the stability of these chemicals in the troposphere, their imperviousness to degradation, their ability to remain in the atmosphere for long periods, the absence of any significant sink, there exists a certainty of a major accumulation of CFCs in the stratosphere. There, by catalytic action of chlorine atoms, photochemical dissociation of ozone molecules is predicted to occur. Such is the lifetime of these chemicals in the stratosphere that even if future releases of CFCs were stopped or severely limited today, ozone depletion would probably still continue well into the next century.

We are dealing then with a problem that has yet to be proven conclusively. It is one that is out of sight, and one that could so easily be out of mind. The problem of ozone is not qualitative but quantitative. Uncontrolled release of chlorofluorocarbons other than CFCs 11 and 12 and of chemicals such as methyl chloroform and carbon tetrachloride, which may also affect the ozone layer, obscure the issue and limit our ability to predict with any confidence the future of the ozone layer. And this problem is compounded by our imperfect understanding of latitudinal, seasonal and other natural variations.

Although the possible consequences of ozone depletion will not be experienced until well into the future, perhaps beyond our own lifetimes, time is not on our side. We have to act now if we are to ensure that the more severe penalties from upsetting the ozone balance are never incurred. These may include consequences for agricultural production, fisheries and human health. Recent research shows that many terrestrial plants, including important crops such as wheat and rice, and aquatic organisms such as fish eggs and larvae, undergo damage when exposed to increased levels of UV-B. The link between exposure to solar UV-B and skin cancer is well established and there are indications that sunlight may also be a causative factor of malignant melanoma.

In retrospect it is impressive how many of the above speculations would be proven to be true: Chemicals causing ozone depletion did indeed include carbon tetrachloride and methyl chloroform, which would be included in the controlled substances list in 1990, and in the same year the suspected effects of UV-B were confirmed by the assessment panels.

The Vienna Convention, March 1985

The negotiations that began in 1984 brought together representatives of the governments of both industrialized and developing nations, spokesmen for the scientific community, and industry representatives, notably those of the Chemical Manufacturers Association (CMA); although NGOs and the media were also present, they did not initially play an important role.

From the beginning, nations formed groups advocating very different approaches. Western nations fell into two factions: the Toronto Group, comprising Canada, the United States, Finland, Norway, Sweden, and Australia and favoring a worldwide ban on the use of CFCs as aerosol propellants but opposing other CFC restrictions; and the EC nations, advocating eventual limits on total production but opposing cuts from current production levels. Developing countries were concerned about the potential of the convention to impose stipulations that might inhibit their own development. Some nations were anxious to address the problem by any means necessary, whereas others—such as the United Kingdom, France, and Japan—were reluctant to agree to any regulatory measures. Among the industrial representatives, the European Council of Chemical Manufacturers' Federations (CEFFIC) took the position that even with significant CFC usage the threat was only a distant possibility and could be met with some future action; in short, they opposed cuts in the production and use of CFCs.

In spite of these disagreements, there was a general acknowledgment that although scientific uncertainties would exist for some time, it was necessary to consider the consequences of awaiting complete certainty, whereas action taken now might prevent irreversible damage to the ozone layer. Not only was cooperation in research, as provided for in the draft convention, necessary, but the potential risks also made it essential to adopt a protocol to reduce CFC emissions. A multioption protocol was advocated that would enable countries in widely differing economic circumstances to accept it while rewarding past actions to reduce CFC use by more affluent nations. The EC proposed a limit on production capacity, reasoning that this would prevent emissions from exceeding the critical value and saying that once the cap was imposed changes in vertical distribution of ozone would be smaller than those caused by an increase in nonaerosol emissions under the multioption approach.

Proponents of the multioption protocol pointed out the shortcomings of the single-option approach put forward by the EC: The production capacity cap was so high compared with current production that irreparable damage to the ozone layer might well occur before it became binding; drastic reductions would be needed when the cap took effect; because no restriction was placed on exports or imports, the protocol would most likely be ineffective overall; and the approach was prejudicial to developing countries. In turn, the EC held that the multioptional approach amounted essentially to a short-term ban on CFCs in aerosols and failed to address the long-term problem, whereas its own approach had been strengthened by steps to secure a substantial reduction in CFCs' current use in aerosols, as well as addressing their use in other fields.

Time was growing short. A plenipotentiary conference was scheduled to convene for the adoption of a convention, and still no agreement on a protocol was in sight. There had even been a proposal to draw up several protocols, or several annexes to the convention, which would deal individually with the various ozone-depleting chemicals. Several governments, particularly those of the industrialized nations, used the lack of scientific evidence and other pretexts as tactics to avoid legally binding obligations on the control of CFCs. In 1983 Sweden, Norway, and Finland presented a proposed draft annex, *Concerning Measures to Control, Limit and Reduce the Use and Emissions of Fully Halogenated Chlorofluorocarbons for the Protection of the Ozone Layer*. The comments on the proposal by a number of nations[1] showed how far apart they were on any agreement regarding controlling ozone-depleting substances.

The draft annex then underwent a number of revisions to make it more flexible and to meet the objections raised by several governments. But major differences among Western countries remained unresolved, and no protocol was drawn up for the consideration of the Vienna plenipotentiary conference in March 1985.

However, before this meeting could take place, news from the scientific community electrified the delegates. Joe Farman of the British Antarctic Survey announced that in comparing measurements taken in the Antarctic spring (September and October) of 1984 with records compiled since the 1950s, he had found that the ozone layer over the continent had been in sharp decline since the late 1970s. Each spring since 1957 the ozone had thinned by about 40 percent; in effect, a "hole" had developed in the ozone layer over the Antarctic as big as the United States and as deep as Mt. Everest, spreading north over Argentina and New Zealand.

Why this should have happened over the Antarctic was not known. It was suggested that the depletion was a natural phenomenon linked to the solar cycle, caused by enhanced levels of nitrogen oxides produced by increased solar activity. Later analyses, however, showed a striking correlation between low concentrations of Antarctic ozone and high concentrations of halogen compounds. Although this was not absolute proof that halogenated source gases (such as CFCs) caused the depletion, it seemed consistent with the hypothesis.

The plenipotentiary conference was convened in Vienna in March 1985 under the auspices of the UNEP and the WMO and adopted a treaty to protect the ozone layer, committing its signatories to take appropriate measures to protect human health and the environment from human activities' potentially adverse effects on the ozone layer. Once again scientific findings had accelerated the negotiation process. The community of nations had decided not to wait for incontrovertible proof of cause and effect, which could come too late to avert irreparable harm, but to take action against a future threat, in what is probably the first application of anticipatory and precautionary principles.

The conference also requested that the UNEP continue work on a protocol on ozone-depleting substances, and the UNEP Governing Council delegated this task to its executive director, at the same time authorizing him "to convene a diplomatic conference, if possible in 1987, for the purpose of adopting such a protocol," clear evidence of the sense of urgency that prevailed.

The Montreal Protocol, September 1987

Negotiations began in 1986 in a changed climate. Scientific information continued to pour in that strengthened the case for an effective protocol, but economic factors had taken on additional importance.

In 1986 the U.S. National Aeronautics and Space Administration (NASA), whose satellites had produced the spectacular infrared maps of the Antarctic ozone hole, estimated that a 3 percent increase in CFC emissions could deplete the ozone layer by about 10 percent by the year 2050; the U.S. Environmental Protection Agency (EPA) said that if CFC emissions continued at recent rates, the resulting extra UV-B radiation could cause 40 million more skin cancer cases and twelve million more eye cataracts over the next century in the United States alone. At about the same time, Swiss scientists reported evidence of an Arctic hole over Spitzbergen, with the ozone layer thinning as far south as Switzerland. The distant threat was clearly coming nearer. Pressure for global regulations and limits on CFC emissions was growing.

Some governments, thinking action premature, resisted, but most governments, most scientists, and some manufacturers, unwilling to gamble with the composition of the atmosphere and the future of the planet, advocated action. Scientists were firmly in favor of controls, without which, they said, changes to the ozone layer would certainly increase cancer and would probably affect agricultural production, atmospheric pollution, and the climate. Before the political and legal negotiations began in Geneva in April 1987, the Toronto Group, led by the United States, changed its position and supported a global freeze on CFC production followed by a series of reductions that would lead to a complete ban. The EC countries wanted a cap on production but no cuts; the Soviet Union and Japan were reluctant to accept any control measures; the developing countries feared that the new protocol would hamper their industrial development.

Trade factors dictated much of the difference between the Toronto Group and the EC: European chemical industries saw U.S. industries as being a step ahead of them in developing substitutes for the CFCs, which could endanger their markets. Because the United States had already banned the use of CFCs in aerosol containers, it was concerned that further controls would affect their use in refrigeration and foam production unless the measures were carefully drafted.

Richard Benedick, head of the U.S. delegation, describes the differences among the delegates:

> The United States and the 12-nation European Community emerged as the principle protagonists in the diplomatic process that culminated in the Montreal Protocol. Despite their shared political, economic, and environmental values, the United States and EC disagreed over almost every issue at every step along the route to Montreal. (Benedick 1991)

The year 1974, when Rowland and Molina made their discovery, was also the year of peak production and use of CFCs, which had been growing at the average annual rate of 13 percent since 1960. The United States was by far the largest producer of CFC 11 and CFC 12, accounting for 46 percent of the reported world total of 813,000 tons. All the EC countries together accounted for 39 percent, with the Federal Republic of Germany the largest producer, followed by France and the United Kingdom tied for second place, then Italy, the Netherlands, and Spain. By 1976 the relative shares were reversed, with EC production accounting for 43 percent of the reported world total and U.S. output having dropped to 40 percent. In 1985, the year before protocol negotiations began, the EC output was 45 percent and U.S. output, which had fallen to nearly half its 1974 peak, was 28 percent of the total. DuPont, in the United States, remained the largest CFC producer, accounting for about a quarter of the world's production, followed by other producers in the United States, the United Kingdom, France, Canada, the Netherlands, Japan, and Latin America.

By 1986 the EC's production share was an estimated 43–45 percent, the United States accounted for about 30 percent, Japan 11–12 percent, and the Soviet Union 9–10 percent. Much smaller amounts were produced by Canada, China, Australia, Brazil, Mexico, Argentina, Venezuela, and India; it was later found that small facilities either existed or were planned in Iraq, Israel, South Africa, South Korea, and one or more eastern European countries. In relation to the gross national product, in the 1980s EC production of CFCs was more than 50 percent higher than that of the United States, with more than half of CFC 11 and 12 sales in the EC made up of aerosol propellants, which had virtually disappeared from the United States.

The EC was clearly the nearly unchallenged CFC supplier to the world, particularly in the growing markets of developing countries. Its exports had risen by 43 percent from 1976 to 1985 and accounted for nearly a third of its production; France alone exported 40 percent of its output. In contrast, the United States consumed almost all the CFCs that it produced. It appeared to the U.S. companies that their European rivals had achieved a competitive advantage in the late 1970s by blocking meaningful regulation; some European countries claimed that the U.S. companies had endorsed CFC controls in order to enter the profitable export markets with secretly developed substitutes:

> Economics certainly played a role in the British position. CFCs were an important foreign exchange earner for the United Kingdom, which exported one-third of its CFC production. Aerosol products accounted for over 80 percent of the country's domestic use of CFCs in 1974, and even in 1986 they still amounted to over 60 percent. One company with a major stake in the production of CFCs and halons—Imperial Chemical Industries (ICI)—influenced U.K. government policy throughout this period. ICI was also the driving force in the European Council of Chemical Manufacturers' Federations, which was an active lobbyist in Brussels and a conspicuous presence during the negotiations. (Benedick 1991)

Fionna McConnel, leader of the British delegation to most of the meetings that negotiated the Montreal Protocol, highlighted the differences between the main producers and consumers of CFCs in a private communication discussing the Benedick volume:

> The book brings vividly back to me the cut and thrust of the negotiations; the cascading volume of scientific evidence, some of it contradictory; confrontations between the defensive chemical industry and the environmental pressure groups; clashes between the EC and the United States; but above all, the sense of shared purpose among the government delegations—even when they were pouring scorn on proposals they did not agree with. . . . If one of the EC delegates had kept a diary, he or she might well have demonstrated that the EC was the true architect of the Montreal Protocol and the United States nearly wrecked agreement because of its obsession with the use of CFCs in aerosols rather than CFC emissions from all sources.

Besides the differences between the Toronto Group and the European Community, the U.S.S.R. was concerned by the lack of substitutes for

CFCs in air conditioning and refrigeration, both essential in the Soviet south, fearing that CFC controls would lead to unrest there. The Japanese worried that no substitutes had been developed for the CFCs used as solvents in their electronics industry, the backbone of their economy. Difficulties also arose regarding halons. Although they were considered serious threats to the ozone layer, a number of countries, chief among them the Soviet Union, were unwilling to impose controls on them because of their use in fire extinguishers in submarines, for which no substitute had been developed.

There were additional problems related to the magnitude and schedule of control measures, production versus consumption, the substances to be covered, and the base year for controls. There were major differences between the negotiators over trade restrictions to be included in the protocol and when it should enter into force, as well as concerns over special treatment of the European Community. The EC was pressing to become a party to the protocol as a separate entity, since its member states had given it authority to act on certain environmental issues. It was ultimately agreed that the EC could become a party, but that it would either cast a single vote on behalf of its twelve members or abstain from voting, leaving its members to cast separate votes.

To summarize, preparations for negotiation of a Protocol on Substances That Deplete the Ozone Layer (later called the Montreal Protocol) began with the following areas of serious disagreement: the Toronto Group countries advocated a production freeze and major cuts; the EC advocated a cap on production but no cuts; the U.S.S.R. and Japan were reluctant to accept any cuts; developing countries feared that any control measures would impede their development; most industry opposed cuts in CFC production and use; and there were differences over the form of a number of points to be included in the protocol.

The formidable task of bringing all these opposing viewpoints into agreement and meeting the 1987 deadline set at Vienna fell to the UNEP. Led by UNEP Executive Director Tolba and Iwona Rummel-Bulska as chief of its Environmental Law and Institutions Unit, a team of experts set to work. The starting place was the issue of control measures, but this was an area of general disagreement where any attempt at negotiation would be doomed to failure. It was decided to hold a workshop rather than a more formal proceeding. Even this was difficult, in that a steering committee of government representatives needed two meetings—in September and December 1985—simply to agree on the organization of such a workshop. The workshop was held in two meetings, May and September 1986. Negotiations on the protocol were to begin the following December; these were completed and the protocol signed by September 1987.

The unexpectedly rapid conclusion of this effort was due to innovations in the negotiation process introduced by Tolba and Rummel-Bulska, who in earlier international negotiations had learned the value of the informal consultation. Away from the scrutiny of colleagues, with comments off the record, delegates can build bridges and form personal relationships that greatly ease tensions and dispel suspicion. This technique proved so useful that it was applied to the negotiation of several other treaties following the Montreal Protocol. But for it to succeed in this case, the scientific community had to provide information that would convince reluctant partners, and industry had to become a part of the process. Another element that speeded the proceedings was public opinion, alerted to the urgency of the situation by the media and the NGOs.

The two workshops, the first hosted by the EC in Rome and the second by the United States in Leesburg, Virginia, were attended by people from government, industry, the United Nations, the science and economics communities, and environmental NGOs, all of whom came, not as representatives of their organizations, but in their personal capacities, to build confidence among the future negotiators and to identify points of both agreement and conflict. Benedick describes the significance of the format: "The idea, launched in Vienna, of convening a special workshop significantly influenced the subsequent history of the ozone issue. In effect, having failed to agree on how to manage the risks, and being well aware of mutual suspicions about the commercial implications of any international controls, the contending parties agreed to step back and together reassess the whole situation." (Benedick 1991, p.)

Although by the time of the Rome workshop new information regarding consumption of CFCs had been provided to the UNEP team by the EPA and independently by DuPont, the workshop did not go smoothly.

Industry representatives from the United States and the EC were still adamantly on opposite sides over the future of CFC production and consumption. It was not until the Leesburg meeting that the participants began to know one another better and the atmosphere became more relaxed. At the second meeting, thirty scientific papers were presented for discussion, covering areas such as demand, commercial perspectives, model calculations, control strategies, cost effectiveness, trade impacts, and ease of implementation. The general position of the scientists was that although scientific uncertainty still existed, much was known, and that there was no doubt that if enough chlorine were put into the atmosphere, ozone would be depleted. Publications and meetings that had occurred between the two workshops had stressed the urgency of the matter and the negative impacts of ozone depletion on health and ecosystems. At the end of the second meeting there was a general agreement that an international regime was needed, and the participants left feeling more ready to begin negotiations. These are the broad areas of agreement that emerged from Leesburg:

• The ozone layer is an exceedingly valuable resource for the entire population of the world.

• It has been, is being, and will continue to be adversely affected by the long-lived chlorine molecules which stem from all CFC products.

• Ozone depletion appears to be more serious the farther away—north and south—one moves from the equator.

• This ozone depletion, by permitting greater quantities of harmful ultraviolet radiation to reach the earth's surface, will pose significant, even if currently difficult to quantify, risks for climate change, for human health, and for ecosystems.

• These risks are considered by virtually all countries [by that time there were recent contributions by the Soviet Union and Japan] as sufficiently serious to warrant control actions.

• Governments and industry are more seriously than ever before considering further regulatory and technological measures to limit future emissions of CFCs.

• The very nature of the ozone layer requires global cooperation if protective measures are to be effective.

This was certainly more than anyone had expected the workshop to achieve, and it augured well for the negotiations that were to begin in Geneva in two months.

The formal negotiating meeting that began on December 1, 1986, was attended by representatives from twenty-six governments, only six of them developing nations, which was an indication of the lack of interest they had shown until then. Four NGOs attended, and industry and business were represented by the European Federation of Chemical Industry Associations and the International Chamber of Commerce. The WMO continued actively to cosponsor the negotiations.

The December meeting was chaired by Winfried Lang of Austria, who had also led much of the work on the Vienna Convention. It considered the fifth version of the draft protocol that had emerged from the Vienna conference, and various versions of elements of the protocol were presented by Canada, the United States, Sweden, Norway, and Finland. Very little progress was made. The chairman presented an information paper on control measures intended to help future sessions focus on the issues. It pointed out that there was no agreement on the scope of the protocol, whether it controlled CFCs only or CFCs and halons; negotiators were equally divided over this issue.

Although the results of the December negotiating session were disappointing, participants still felt there must be a common global endeavor to deal with the threat of ozone depletion. There was at least clear agreement that Article II of the protocol, which dealt with control measures, should be equitable, easy to implement, flexible enough that it could be changed as new technological developments and scientific evidence emerged, and responsive to as many situations as possible, and that it must acknowledge the needs of developing nations.

New problems emerged during this first phase of negotiations, some of which were very difficult. Some installations for the production of ozone-depleting substances had been begun and were yet to be finished, an issue raised by the Soviet Union and echoed by South Korea. The developing countries wanted to clarify the nature of the financial contributions required to implement the treaty. There were serious disagreements about the provisions for the treaty to enter into force, primarily between several groups: between developed and developing countries; between developed countries that produced CFCs and those that did not; and especially between the EC and the United States. The special position of the EC was questioned, as were details of how to assure compliance.

Article II at the time of the meeting contained only five paragraphs, but the measure of disagreement on those five could be seen by the fact that in them there existed wording enclosed within eighteen sets of brackets, the signal in the negotiation of legal instruments that the bracketed item cannot be approved as it stands.

Less than three months later, in February 1987, the negotiators met again. This time 31 governments sent representatives, ten of them developing countries, and more NGOs attended; the attendance of representatives from industry and business was largely unchanged. It should be noted that the participation of delegations from industry, business, and NGOs in environmental treaty negotiations, once unwelcome, is now encouraged, although the U.S. representative at this meeting remarked that many in the United States felt that some other nations were more concerned with short-term economic gains than with the well-being of future generations.

At the start of the meeting the United States reiterated its position: a freeze on adjusted CFC production, then reductions up to 95 percent followed by a review process. The EC agreed to an early production freeze, a total ban on imports from nonsignatory countries, and the possible desirability of some reduction in production. Japan agreed on controls of two CFCs, CFC 11 and CFC 12, but was reluctant to endorse the rest. The developing countries said they could not be expected to control emissions to the same extent as developed countries. The NGOs wanted a rapid phaseout, to be nearly complete within ten years. The UNEP executive director suggested a 20 percent reduction every other year beginning in 1990, resulting in an almost complete phaseout by the year 2000.

So four difficult issues presented themselves: agreement on the status of science; control measures; trade provisions; and the special situation of developing countries. Four working groups were chosen to deal with them. The two most problematic were control measures and trade restrictions. The debates also brought out issues that resulted in the establishment of very important principles:

• Contributions to financing implementation of the protocol should be commensurate with the amount of CFCs produced and used. From this simple reasoning grew two powerful principles: burden sharing and common but differentiated responsibility.

• The protocol should ensure that new technologies and substitute chemicals are not denied to developing nations, which also addressed an ongoing issue, technology transfer as it relates to patents and intellectual property rights.
• Imports should be restricted from nations not parties to the protocol, and movement discouraged of capital and facilities outside the protocol areas, a clear exception to GATT and its provisions at that time being negotiated in the Uruguay Round of GATT for Trade Liberalization.
• A study was required of the socioeconomic consequences of the control measures.

Added to these issues in the debate was one that had remained high on the international agenda since negotiations began on the first International Development Strategy of the United Nations, transfer of financial resources from North to South. It began to appear that the protocol could not come into force until at least six or seven years after adoption, and that reductions of CFC production and use would not begin for a decade.

The February 1987 negotiations produced a sixth draft protocol, with an article on trade made up of six paragraphs with ten brackets and one on control measures having four paragraphs with fourteen brackets. However, by the time the negotiators met again in April several significant changes had occurred. Scientists had agreed almost unanimously that ozone was in the process of being depleted; the NGOs had begun to take a major interest in the issue; the media were demanding action; and much of industry, having seen that international controls were inevitable on ozone-depleting substances, had decided to cooperate. Repeating a tactic that had proven valuable in earlier negotiations, Tolba and Rummel-Bulska had also begun to hold informal consultations with the government representatives.

To quote Benedick once more: "UNEP Executive Director Mostafa Tolba set the tone with an impressive opening address. Referring to the work of the Wurzburg meeting [the Scientific Models Meeting], Tolba emphasized that 'no longer can those who oppose action to regulate CFC releases hide behind scientific dissent.' In the face of the potential of CFCs to cause unmeasurable damage to our planet, he unequivocally placed UNEP behind tough international regulations. From that point on, Tolba assumed a central role in the protocol negotiations, exerting his personal influence and his considerable authority as scientist and head of a UN organization" (Benedick 1991).

Tolba and Lang, chairman of the negotiating group, organized a small closed group of key delegation heads, called "Friends of the Chairman," to meet informally, primarily on the crucial issue of control measures. This group consisted of heads of delegations from Canada, Japan, New Zealand, Norway, the Soviet Union, the United States, and the EC, as well as the current, past, and next president-nations of the EC (Belgium, Denmark, and the United Kingdom). Members were chosen to reflect a country's weight in the CFC market and its interest in the ozone issue and to achieve a geographic balance. Developing nations were unrepresented not because of a lack of interest, but because they were satisfied with an article that had given them a ten-year grace period for compliance, and acknowledged their need for technology transfer, substitute chemicals, and financial resources. Trade restrictions had been drafted by a working group led by Ambassador Hawas of Egypt, to ensure that no arbitary nontariff barriers existed under the guise of environmental protection.

Working away from the plenary sessions and on an unofficial text, the representatives of this group were able to be more flexible, because they were not committing their governments to any course of action. At the end of the session the group produced an unofficial draft (labeled "Tolba's personal text") in which many governments expressed interest. The response was so positive that the press erroneously reported that an agreement had been reached. However, the differences between the EC and a number of other countries remained significant.

In June this informal group was reconvened in Brussels for more work on control and other major provisions; in July a small number of legal experts met in The Hague to analyze the entire protocol and produce a relatively consistent draft for the final negotiating session in Montreal. The working groups continued the struggle to resolve the remaining issues.

Meanwhile the 1987 Montreal session loomed: the plenipotentiary conference was to take place from September 14 to 16, preceded by five days of preparation by the working groups. More than sixty governments would attend, mostly at the ministerial level, of which more than thirty were of developing nations. Industry, NGOs, and the international media were also to attend, and yet there was still no protocol.

The Friends of the Chairman took up the most intractable issues: control measures, new installations, entry into force, and special treatment of the EC. Beginning September 9 they met daily in closed informal sessions, sometimes well past midnight. Compromise proposals were offered, simply to test their wording against the concerns of the delegates; nerves frayed and tempers were sometimes lost, especially in matters affecting the EC. Still there was no protocol.

As the plenipotentiary meeting was about to open, Tolba agreed with Lang, the conference chairman, on a strategy. Each would make an opening statement and immediately, before the ministers and heads of delegations could establish positions they might find it hard later to abandon, the chairman would adjourn the meeting. This was done, and there was a long hiatus, the ministers occupying themselves as best they could in the corridors of the meeting hall while behind closed doors the informal consultations continued at fever pitch. As time went by, it became very embarrassing that nothing was emerging, even though in fact three of the issues had by the afternoon of the second day been resolved: control measures, new installations, and entry into force. The main problem, the special treatment of the EC, seemed insoluble. At last Benedick and Laurens Jan Brinkhorst, the Director General of Environment of the EC, reached an acceptable compromise, which required Brinkhorst to obtain agreement from the twelve EC countries.

This was at 4:00 in the afternoon of September 15, and the ministers were leaving the hall to prepare for an evening reception given by the mayor of Montreal. Brinkhorst met with strong resistance from some of the representatives, and three had to be called by their governments and instructed to join the consensus. At 5:00 the language was agreed on. The small informal group was hurriedly reconvened and accepted all the negotiated texts. When Tolba and Rummel-Bulska appeared at the reception and announced that they had a treaty, they were met with general amazed disbelief.

Briefly, the control measures issue was resolved by a compromise that would reduce production and consumption of all five CFC types by 50 percent by 1999, using 1986 as the base year. There were a number of small adjustments to be made, and language was found to meet the situation of the U.S.S.R.[2] by allowing new installations to produce CFCs if they had been approved in national legislation before January 1, 1987, and if they were completed before December 31, 1990. On the matter of entry into force, the aim was to assure that at the time the protocol went into force, the major producers would be signatories. (The proposal that at least 75 percent of CFC production be by countries ratifying the protocol before it entered into force was categorically rejected by developing countries, and by developed countries that did not produce CFCs, who said it turned the protocol into a producers club.) The unanimously accepted version required that at least two-thirds of the global consumption of the substances be by parties to the protocol before it entered into force. This ensured the membership of the United States, Japan, the EC, and the Soviet Union, as well as some smaller consumers.

Reaching agreement on the wording of the protocol, however, was not the end of the difficulties. The group of government experts who were in charge of assuring that all the documents were legally correct felt that the language was clumsy; there was nothing wrong legally, they said, but they wished the document to flow more smoothly. Tolba and Rummel-Bulska, working with them far into the night, finally convinced them that any change in the language that had required such delicate negotiations to achieve would destroy the compromise. Somewhere after midnight the experts agreed, and the race began to get the document translated, printed, bound, and ready for adoption and signature in time for the final day's meeting—early in that day, in fact, because many of the ministers would leave during it. Largely due to the efforts of Rummel-Bulska, the deadline was met and the signatures were obtained.

The feeling of triumph was general. This was the first truly global environmental treaty, and moreover it dealt with an issue still shrouded in scientific uncertainties, one that posed a threat, not immediately, but in the future, one that potentially affected everyone on earth today and far into the future. It was a monument of collective action, a masterpiece of compromise. It had the advantages of ease of implementation, flexibility due to its mechanism permitting adjustments to meet scientific, technological, and socioeconomic changes, and the clearly applied principle of common but differentiated responsibility. It was also the first treaty to set for itself, subject to conditions, a date for its own entry into force: January 1, 1989, barely fifteen months after the treaty had been signed. There would follow a number of meetings to finalize its details, but the date—September 16, 1987—went down as a landmark in the history of international negotiation.

The Road to Helsinki

Six months after the Protocol was signed a major report was released by the NASA-NOAA Ozone Trends Panel, demonstrating conclusively that human activities were causing atmospheric concentrations of chlorine to increase around the world. Even more disturbing was the finding that from 1969 to 1986, a small but significant depletion of the ozone layer, amounting to 1.7 to 3.0 percent depending on latitude, had already occurred over heavily populated regions of the Northern Hemisphere including North America, Europe, the Soviet Union, China, and Japan. The announcement made headlines around the world. In response the four assessment panels called for by the Protocol were convened early by the UNEP leadership, even though they had yet to be authorized by the First Conference of the Parties. This unconventional proceeding would later be endorsed by the parties, who agreed that action was urgent in the face of the new scientific evidence.

The governments of the United States, the Netherlands, Germany, and the EC covered the cost of the panels, which consisted of some six hundred high-level scientists, economists, biologists, medical scientists, engineers, and others, from all corners of the world. Working long days, often into the night, their foremost conclusion was that control measures had to be tightened. The Protocol articles covering finance and technology transfer were more difficult, and informal consultations were again resorted to. Media pressure for more concrete action was intense.

Two months after the entry into force of the Protocol, and only a month before its contracting parties were to meet in Helsinki, British Prime Minister Margaret Thatcher convened an international ozone conference in London, in March 1989. Although this appeared to many to be a preemption of the Helsinki conference, it turned out to be very helpful. More than 120 governments participated in the London meeting, more than two-thirds of them at the ministerial level, a far cry from the meager two dozen governments represented at the start of negotiations and the sixty in Montreal. More than ninety environmental organizations, as well as media from all over the world were there, all calling for a stern approach to the problem. Several countries announced in London that they would ratify the Protocol; developing countries reiterated their demand for assurances that financial resources and technology transfer would be forthcoming. Tolba, who also attended the London conference, stressed that "as the industrialized nations begin to address the interlinked threats from ozone depletion and climate change, the developing nations [must be] assisted and not inhibited in improving their economies. . . . There is a need for international mechanisms to compensate them for forgoing the use of CFCs. . . ."

Helsinki, April 1989

The deluge of scientific information that had preceded the opening of the Helsinki conference assured that stronger control measures would be enacted. The difficult areas were finance and technology transfer. China, supported by India and other developing countries, had proposed establishment of a special fund for the Protocol. Neither China nor India was a party to the Protocol; in fact only thirty-one of the contracting parties were in Helsinki, but they were accompanied by fifty-one nonparties, which showed the global importance attached to this issue.

On technology and compliance, the following elements were chosen as components of the work plans required by the Protocol:

1. dissemination of the scientific, environmental, technical, and economic assessment panel reports, and of the synthesis report, and their follow-up
2. regular updates, particularly regarding environmentally sound substitutes or technological alternatives to the use of CFCs or halons
3. production and wide dissemination of informational material to the public
4. development of methods for exchange and transfer of environmentally sound substitutes and alternative technologies
5. establishment of liaisons with international organizations and financing agencies that could help implement the Protocol

6. establishment of an open-ended, ad hoc working group of legal experts to develop and submit to the Secretariat by November 1, 1989, procedures for determining noncompliance and actions to be taken when proven

On the issue of financial aid to developing countries it was decided (a) to recognize the urgent need to establish international financial mechanisms to help developing nations comply with the Protocol both at present and as it would in future be strengthened; and (b) to establish an open-ended working group to develop such mechanisms, including international funding mechanisms and not excluding the possibility of establishing an international fund, its findings to be reported at the second 1990 meeting.

The next Conference of the Parties was to be held in London in May 1990. The intervening thirteen months were filled with meetings, negotiations, and consultations, and every known kind of negotiating tactic was used in the search for agreement. The three most pressing issues were the need for a revised set of control measures aiming at an early phaseout of controlled substances and the inclusion in the Protocol of other ozone-depleting substances; finances; and technology transfer. The three working groups that formed were made up largely of people who by this time knew each other very well and were determined to reach agreement.[3] The formula of informal consultations among them, between two or more individuals, between two or three countries or larger groups depending upon the topic, would make possible a clear understanding of the real issues that underlay their differences.

It became evident to the working groups that, successful and widely hailed as the Montreal Protocol had been, in light of what had become known since its inception it was now inadequate. Had the makers of the Montreal document known what was known at Helsinki—that CFCs are major greenhouse gases threatening to produce perhaps 20 percent of global warming—its regulations would doubtless have been more stringent, its timetable for control more urgent.

Finances, especially as they related to technology transfer, were an urgent problem. China and India, who had suggested an international fund, had the potential to undo all the work of the Protocol by their production of ozone-depleting substances unless alternative technologies were made available to them and other smaller producers. To their credit, developed countries (except, for a brief period in 1990, the United States) saw the advantages to be gained from such a fund. They agreed, and then negotiators considered what it would entail. What should it cover? How large should it be? Who will pay? How will it be disbursed? Who will manage it?

The Second Conference of the Parties, in London, was a scant ten months away, and the working groups broke up into sub-groups: on finances, chaired by Ilkka Ristmaaki (Finland), on programs, chaired by Juan Antonio Mateos (Mexico), and on control measures, chaired by Victor Buxton (Canada). The groups met in Nairobi, August 21–25, 1989. Clearly the negotiations were polarized between the developed and the developing nations, the latter considering themselves to be victims of a deteriorating environment and considering the former to be the cause of the problem. The reticence of the developing countries to ratify the Protocol was born, not of a lack of concern, but rather of a lack of the resources necessary to comply without disrupting their development efforts. Yet in spite of the potential for politicization, all the negotiators abandoned dogma to seek solutions together, and all were willing to give and take freely in search of an agreement.

Before the Nairobi meeting the financial subgroup met in informal consultation in Geneva with a small number of financial experts to discuss various possible institutional arrangements. Out of the several that were discussed, ultimately it was decided to establish an international trust fund or other effective mechanism to which donor countries would pledge contributions and which would be serviced by a financial and technical secretariat. Donors were to be official government sources in developed countries, and recipients were to be developing countries that were parties to the Protocol. Other stipulations included burden sharing among donor nations and the requirement that funds be applied as compensation for the costs of the transition from ozone-depleting substances, which would allow developing countries to comply with the Protocol.

Agreement on a trust fund was complicated by the fact that not enough information was available to determine how large it should be. Feasibility studies were proposed and widely supported; the developing countries felt that because of the need to proceed with setting up funding mechanisms, currently available estimates should be used to proceed pending more accurate data. It was agreed that work would progress on both issues simultaneously. With the decision that a trust fund would be established, the working group turned its attention to the matter of its magnitude. It was felt that studies on technology options and their cost should be carried out in developing countries that manufactured or imported the various categories of ozone-depleting substances, and that the results of these studies should be made available to the next working group meeting and to the June 1990 Conference of the Parties. The study would also examine options for an international financial mechanism, especially those proposed by the UNEP executive director: a specific new fund, an international environmental facility, a pilot investment program (ECOVEST), or an international corporation. It would study the potential for using existing institutions, either separately or in combination, including UNDP, UNEP, UNIDO, the World Bank, and regional banks.

During the period before the Nairobi conference, the many options had been narrowed to those that had the widest support, but it was found that the issues were still too complex to solve in only five days. Another meeting of the working group was held from August 28 to September 5 to tighten the control measures of the Protocol, add new chemicals to the list of controlled substances, and clarify a number of other issues. By the end of the meeting a bracketed text had been prepared for presentation at the next London meeting, in compliance with the rule that requires six months to pass before any proposed changes can be considered by a Conference of the Parties. Negotiations on the text continued throughout the nine months before the June 1990 meeting, starting in September 1989, with informal consultations preceding each of a series of meetings in February, March, May, and June.

The developing countries were concerned that if the proposed early phaseout of the controlled substances was enacted, they would become not only scarce but expensive during the ten-year grace period. However, at the urging of Tolba, they found common grounds for agreement with the industrialized nations. Addressing the group in September 1989, Tolba pointed out that the Montreal Protocol was inadequate and that immediate action must be taken to achieve "nothing less than the complete phase-out of all fully halogenated CFCs, and control [of] halons, carbon tetrachloride and methyl chloroform in preparation for their eventual elimination."

In addition to the timing of control measures, there was intense informal discussion of the financial mechanism to be employed and of the problems of patents and proprietary rights in the substitute chemicals that would make up a technology transfer. Most difficult was to work out the paragraph that tied the obligations of developing countries to comply with the Protocol to the latter two issues. This would in fact only be resolved during the final hours of the London conference.

While remaining in close contact with the working groups, Tolba brought all the relevant stakeholders into the discussions by arranging for a meeting of the technology transfer subgroup with the director-general of the World Intellectual Property Organization (WIPO) and contacting the secretary-general of the International Chamber of Commerce to find out how the industrial conglomerates saw the issue. He also suggested a new paragraph that would make compliance with the phaseout schedule conditional on availability of financial assistance and technology transfer, and pointed out that technology transfer should be on a preferential and noncommercial basis.

At the May meeting of the working group, Tolba presented a proposed text covering the funding mechanism that, because of its length and complexity, can only be summarized here. It contained seven items:

1. The funding mechanism would provide financing to facilitate compliance by developing countries through a multilateral fund that would act as a clearinghouse and would accept contributions in kind.
2. It would finance all agreed incremental costs of the countries operating under Article 5 of the Montreal Protocol (essentially, developing countries).
3. Contributions would be based on a scale using 1986 as a baseline year.
4. Contributions would be in addition to other funding received by the recipient countries.
5. The fund would operate under the authority of an executive committee with a balanced representation of developing and developed countries.
6. To encourage bilateral cooperation between developed and developing parties, contributions in the form of bilateral cooperation, in cash or in kind, could be deducted from agreed contributions to the multilateral fund.

7. Financial transfers would be made through the clearinghouse function of the fund.

Ten days of negotiations preceded the Conference of the Parties in June. They yielded very little with respect to technology transfer and financing beyond an agreement to establish a fund and defining what it should finance. Details of its management remained unresolved, and the developing countries, particularly those of Asia, were concerned about the availability of CFCs or their substitutes. In spite of nearly round-the-clock meetings of the working group and the efforts of its chairmen, especially Ristmaaki and Mateos, and despite continuous informal consultations held by Tolba and Rummel-Bulska, by the end of the preparatory meeting there was still no agreement on the issues of the financial mechanism and technology transfer.

London, June 1990

The Second Conference of the Parties was attended, mostly at the ministerial level, by fifty-four Parties and forty-four nonparties. All the delegations were headed by ministers of environment except for that of France, led by the minister of foreign affairs. Thirty-four industrial groups and fourteen environmental groups were represented.[4]

The Secretary of the United Kingdom's Department of Environment, Chris Patten (later to be the last British governor of Hong Kong), was elected president of the meeting. He and his colleague, United Kingdom Minister of Environment David Trippier, worked around the clock with Tolba and Rummel-Bulska to overcome a host of obstacles. Some delegates occupied entrenched positions from which they had no intention of moving, and it took the combined efforts of many delegates to make progress.[5]

The UNEP Secretariat continued to play an active role, mostly through Tolba and Rummel-Bulska, whose conviction of the urgency of their mission never faltered and is indicated by Tolba's address at the opening of the conference:

> It is a source of the most profound encouragement to UNEP that we meet here in London with contracting parties now indicating readiness to provide those incentives. I refer to a financial mechanism and to technology transfer.

> What we must agree upon is the establishment of a financial mechanism including a properly financed multilateral fund, one designed specifically to meet the incremental costs imposed on developing nations when complying with the provisions of a current and strengthened protocol. . . . What remains is the last step. It is your responsibility to cover it. It may be difficult—but it must be covered.
>
> A successful outcome will demonstrate to a skeptical audience worldwide that the nations of the industrialized North are serious about tackling the inequity in the global economy which is the underlying reason why our human environment is being destroyed.
>
> Any setback here in London would amount to a setback for a whole movement to save the environment, a setback from which the world will never recover. I am sure we all bear this in mind as we get down to the task at hand.

The meeting was scheduled for three days. For the first two days, while the open negotiations went on without much progress, behind the podium in a tiny room, Patten, Trippier, and Tolba held informal consultations with a stream of individuals and groups, with handwritten compromise formulations proposed and discussed, while Rummel-Bulska monitored the plenary meeting, chaired by its vice presidents. Now and then one of the behind-the-scenes players would rush to the conference room to consult with key delegates over some new compromise language. By the end of the second day this strategy had produced some tenuous agreements, and Patten consulted informally with the heads of delegations, beginning in the morning of the final conference day and continuing until evening. Then the meeting became open, and within an hour all the decisions agreed to in this manner had been formally adopted.

So against all odds, the London meeting succeeded. Governments agreed to phase out CFCs by the end of the century and to include a series of other chemicals in the list of controlled substances. They agreed to establish a special fund to help developing countries comply with the Protocol and devised rules for contributions by industrialized countries as well as a management plan. Language was found to provide developing countries a grace period for coming into compliance and to allow the transfer of alternative chemicals and technologies.

These amendments to the Montreal Protocol required ratification, and ratification requires time. So that valuable time would not be lost, the parties agreed by a simple decision to establish an interim fund of $240 million for 1990–1993, with allocations to be determined by size of country:

$5 million to a small country, $20 million to a medium-sized country, and $40 million to a big country, such as China or India. At the time this was decided, China and India had not become parties to the Protocol; these financial incentives led China to join in 1991 and India to follow soon after. The Canadian Minister of Environment, Robert de Cortet, offered Montreal as the seat for the new fund, beginning with the interim fund, at no cost. The offer was accepted and the offices established on January 1, 1991.

Copenhagen, November 1992

By the time of the next major meeting of the contracting parties, it had become apparent that more stringent controls were needed, and new compounds had been suggested for the control list. One significant chemical, methyl bromide, was important to Israel, which exported fruit treated with methyl bromide sprays and used the chemical as a soil fumigant. Israel and a number of developing countries led a campaign against its inclusion.

During the twenty-eight months that had elapsed between the London and Copenhagen meetings, two important conventions were drawn up, one on global climate changes and one on biodiversity. They were supported by the Global Environment Facility (GEF), which had basically the same structure as the Montreal Fund—World Bank, UNDP, UNEP—and were to support action in the areas of climate change, international waters, biodiversity, and ozone layer protection. The UNEP wanted to convert the Montreal Fund, still an interim fund at that point, into an established fund as had been planned; France and the United Kingdom wanted to integrate it into the GEF. Before the Copenhagen meeting, Tolba paid an official visit to France, meeting with representatives of the ministries of environment, foreign affairs, and finance and pointing out that if the permanent fund were not established developing countries might disengage from the Montreal Protocol.

At the Copenhagen meeting, the familiar pattern reappeared of working groups, subgroups, and informal consultations. With the presence of a large number of negotiators who had established friendly relations, all working toward a common goal and guided by the Minister of Environment of Denmark and Tolba, the two most difficult issues were agreed upon at the eleventh hour: more stringent control measures and establishment of a permanent fund.

Lessons from the Montreal Protocol

The sense of urgency shared by the negotiators of the Montreal Protocol not only led to the use of informal means to conclude an international treaty, in itself an innovative process, but also resulted in the establishment of a number of precedents. One was that once a two-thirds majority of the delegates had approved the adjusted control measures, they became binding on its signatories without the lengthy process of formal ratification. Another was the establishment of the Interim Multilateral Fund by a simple decision of the Conference of the Parties, although there was no legal basis for the step in either the Framework Convention or the Protocol; the fund's establishment under such circumstances was considered by some to be an unprecedented step. The noncompliance provisions were similarly unprecedented. Other decisions that were made by consensus, which would have under normal legislative rules required years of formal negotiation, included the requirement that any production increase of controlled substances be on the condition that they not be for export. The data-reporting provisions were similarly streamlined to permit swift action.

We know the dangers of ozone depletion are still with us. The chemicals that have been substituted for CFCs may themselves contribute to global warming. But we can be encouraged by the knowledge that the years 1982–1992 launched a new kind of diplomacy that can be aptly called global environmental diplomacy. Asked to comment on the Vienna Convention and Montreal Protocol and their subsequent modifications, Eileen Claussen, one of the most effective members of the U.S. delegations and later the U.S. Assistant Secretary of State for Environment, said:

> I view the Montreal Protocol as perhaps the best example we have of an effective international environmental agreement. . . . [T]here were four key ingredients which made a difference time and again:
>
> 1. *A core group of countries.* Starting pre-Montreal, there was a key group of countries intent on moving to a Protocol dealing with CFCs, including the CANZ

countries (Canada, Australia and New Zealand), the Nordic countries (Sweden, Norway, Denmark, Finland), other EFTA countries (Austria and Switzerland), and the U.S. Without a continual push from those countries, I doubt that the Protocol would have been signed in Montreal. In the Montreal to London period, it seems to me that the most difficult issue to deal with was the establishment of the Fund. And here again, there was a core group of countries that played a major role (Mexico, the Netherlands, the Nordics, the U.S.—save for that one difficult period where the White House played hardball). From London to Copenhagen, the core group on the Fund prevailed again. The control issues were relatively straightforward, except that we did not have a core group of countries on methyl bromide, and we didn't achieve much either.
2. *The role of science and technology*. I think it is fair to say that science and the consensus among scientists around the world was a critical ingredient in the Protocol process, as was technology, and the consensus that emerged on what could be accomplished and by when. Even more important, I think, was the assessment/reassessment process in the Protocol, since this forced a review and was almost impossible for the Parties to discount. In fact, without assessments for both science and technology, I doubt that we would have been able to move either to the London or the Copenhagen agreements.
3. *Willingness to compromise*. While it is certainly true that the Montreal Protocol involved many countries with strong views on what should be done, there was always a willingness to take one step at a time. If we couldn't agree on a phase out for CFCs in 1987, we accepted 50%. If we couldn't agree to a methyl chloroform phase out by 2000 in London, we accepted 2003. If we couldn't reach an agreement on a phase down schedule for methyl bromide in Copenhagen, we agreed to freeze. Of course, countries agreed to the interim steps because they believed they could come back another time and make the step more stringent (which is . . . what happened at every stage). But I think the spirit of compromise was critical to the success of the Protocol.
4. Some *strong personalities*. In the end, it all boils down to individuals and personalities. And the Montreal Protocol had more than its share of strong and effective ones. Could we have achieved what we did without [Mostafa Tolba]? Absolutely not. Someone had to pull all the parts together, know who to ask to do what, cajole, coax, push and pull. Would we have a fund without Mateos? It was key that the G-77 had leadership that was savvy about when and where to take a stand, and when and where to compromise. I believe he was critical in both London and Copenhagen. Could we have moved the phase-out dates without Per, Vic and all the Steves? I think not. Did [I] make a difference? I think so. So I am convinced that, above all, successful negotiations require strong and effective participants, and while I cannot speak to all the other Conventions, my guess is that the best of them had strong players as well.

To this assessment must be added the lessons learned during the long course of the negotiations. First, the sine qua non of environmental negotiation is the mobilization of public opinion. The Vienna Convention was agreed to in a climate of only mild interest: Neither the NGOs nor the media had drawn enough attention to the ozone problem to arouse the public. On the other hand, when reputable scientists reported that the ozone layer was being eroded above the Antarctic and that the likely results would be increased risk of cancer, cataracts, and crop failures and reduced immunity, the media responded with headlines and an aroused public pressed for quick action. Citizen groups and NGOs demanded and got swift negotiation, adoption, and the entry into force of a viable control mechanism, all in the space of six years, December 1986 to December 1992. From scientists' first signaling the role of CFCs in ozone depletion in 1974 to the Framework Convention's entry into force in 1989 had taken fifteen years; it took only two years to negotiate, adopt, sign, ratify, and enforce the Montreal Protocol. Scientific certainty mobilizing public concern made the difference.

Tolba and Rummel-Bulska, representing the UNEP, abandoned the traditional mediator's role of noncommittal neutrality and actively sought solutions, taking the position that as citizens of the world like everyone else they had a stake in the outcome, using their skills to bring about compromises and finding sound legal language to express them while avoiding the appearance of favoring one party over another. This was an important factor in the successful outcome of the negotiations.

An important principle was developed and crystallized in the Montreal Protocol and its amendments, that of common but differentiated responsibility. Although it had existed since the 1972 Stockholm Conference, it had not previously been incorporated into a regulatory framework. This was a strong encouragement to the developing countries to remain wholeheartedly in the negotiating process.

Probably the most important lesson of the ozone treaty negotiations was the value of the informal consultation. Away from the negotiating table, although the negotiators' goals remain the same as in the formal sessions, because they are not committing their governments in these off-the-record conversations they can be more relaxed and more open to seeing the others' interest in reaching common solutions and to making compromises. They become friends working for a common cause. Following this successful formula, the UNEP has employed informal consultations in other treaties: the Basel Convention on the Control of Transboundary Movement of

Hazardous Wastes and Their Disposal, the preparatory meetings for the Climate Change Convention and the Biodiversity Convention.

None of these factors, however, would have been enough to bring about these international environmental agreements had it not been for the basic change in national attitudes away from the primacy of national sovereignty and toward international cooperation. During the years of negotiations the issue of sovereignty was never raised: Each country was in a sense interfering in the internal affairs of all the others in safeguarding the environment of all.

There is still at this writing some work remaining to achieve the goal of protecting the ozone layer. Although governments have been unwilling to strengthen the noncompliance procedures and to carry them out vigorously, it can safely be said that the legal obligations established under the Ozone Convention and Protocol are preventive rather than remedial. There is also cause for optimism: The two instruments were designed to be flexible and adaptable to changing conditions, and the mandated periodic assessments will assure their relevance and usefulness well into the future.

Notes

1. *Australia*: "We have no particular comments to make at this stage on the proposed annex concerning regulation and control of CFCs submitted by Finland, Norway and Sweden. The relevant issues would be better covered in a protocol than an annex."
Belgium: "The draft text of article 1, in proposing termination of the use of CFC 11 and CFC 12 for non-essential uses, goes too far. We fear that the definition of essential uses will differ from country to country, which may cause distortions in trade between the various countries."

Canada: "Draft Article 2: In general, Canada is of the view that controlling such non-aerosol uses of CFCs is premature but could be subject of a future annex or protocol. Canada could not sign an annex or protocol with such an article since it would require significant changes in domestic regulations which in Canada's view are not justifiable at present."

Denmark: "The Danish Government considers the proposal to be a very good one."

Italy: "Any regulation concerning chlorofluorocarbons should be adopted exclusively through legal means to be eventually stipulated after the signing of the convention."

Japan: "The Government of Japan is of the opinion that, at present, the fact of changes in the ozone layer, identification of the substances causing such change, and the mechanism of destruction of the ozone layer have not yet been scientifically established. It is, therefore, not appropriate to impose on nations any legal obligation by this annex constituting an integral part of the Convention."

Madagascar: "The Malagasy authorities concerned have no objection to the draft proposed by Finland, Norway and Sweden."

Netherlands: Suggested a number of modifications to two draft articles.

New Zealand: "Draft article 1 is generally in accord with New Zealand Government policy, although it may be difficult to fix a target date because of the unreasonable hardship this might impose on a few small users of CFCs. Draft article 2 is considered to be of doubtful value as we are unaware of any practicable technologies, existing or foreseeable, which could be used to limit emissions from foam plastic or refrigerators (at the end of their useful lifetimes). Article 3 is regarded as feasible."

Switzerland: "In our view, the convention should contain in addition to general provisions specific regulations relating to CFCs. These should be dealt with in an additional protocol to the convention which would be binding on all parties to the convention itself, and would be distinguished from the annexes to the convention which should remain more technical in nature."

Thailand: "Chlorofluorocarbons (CFCs) as used in aerosol products were included in the list of poisonous substances which are banned from import, manufacture and sale in Thailand."

United Kingdom of Great Britain and Northern Ireland: "During negotiations in the Ad Hoc Group the United Kingdom Government has placed a general reserve in regard to all possible annexes and protocols on the grounds that they are premature until a framework convention has been elaborated. The United Kingdom Government believes the Nordic proposal is unnecessary and unsound. Any eventual protocol should go no further than the position of the European Community which has agreed on a temporary reduction of production subject to review. The United Kingdom Government hopes, in the light of the above, that the proposers of the draft annex will be persuaded to withdraw."

2. The period of these negotiations saw the breakup of the Soviet Union, but because its negotiators were from Russia, the location of its outmoded, inefficient industries, their participation continued unchanged.

3. Tolba and Rummel-Bulska were personal friends of most of the key negotiators. Among those to whom credit goes for bringing the negotiations to a successful conclusion were Ilkka Ristmaaki (Finland), Ines Schuidtziara (Germany), Juan Antonio Mateos (Mexico), Eileen Claussen and Steve Anderson (United States), Victor Buxton (Canada), Per Bakken (Norway) and Steve Lebabti, Fionna McConnel, and Patrick Szell (United Kingdom).

4. Some of the delegations were quite large: Canada (17), EC (18), Germany (13), Japan (18), Spain (13), the Netherlands (18), the United Kingdom (28), the United States (20), China (12), and the Republic of Korea (10). Environmental

groups included Friends of the Earth International (8), Greenpeace International (8), and the Australian Conservation Foundation (11, including a youth delegate). Japanese industry was represented by four associations with twelve representatives, and industry representatives attended from the United States, the United Kingdom, France, Austria, Belgium, Switzerland, the Netherlands, Korea, Taiwan, and India, with the largest representation from the United States and the United Kingdom.

5. Among the key players throughout the meeting were Maneka Gandhi, then the Minister of Environment for India; Wang Zangzu, then the Deputy Chairman of the Chinese National Environment Protection Agency; and William Reiley, then the administrator of the U.S. EPA. Among the many others who played important roles in working toward the common goal were John Whitelaw (Australia), Willy Kempl (Austria), Laurens Jan Brinkhorst (EC), Kay Barlund and Aira Kalela (Finland), Brice Lalond (France), Klaus Toepfer, Otto Vaeland, and Ines Schuidtziara (Germany), Franciska Issaka (Ghana), Ikeda, Kajima, and Takahashi (Japan), Tan Meng Lang (Malaysia), Juan Antonio Mateos (Mexico), Hans Alders, Willem Kakebeke and Hans Lammers (the Netherlands), Philip Woollaston (New Zealand), Birgitta Dahl (Sweden), and Flavio Cotti (Switzerland).

6

Facing Up to Global Warming

Whereas the damage to the ozone layer from industrial chemicals was discovered only relatively recently, and action taken relatively quickly, the problem of global warming due to human activities has proven much less tractable. We have known since early in the twentieth century that a buildup of carbon dioxide (CO_2) in the atmosphere could affect the world's climate; by the late 1960s, it had been confirmed that the CO_2 concentration in the atmosphere was rising, largely because of human activities. The Agricultural and Industrial Revolutions profoundly changed the rate at which humans were using fossil fuels and burning off the world's forests, releasing CO_2 into the atmosphere at an unprecedented rate.

The world's climate has naturally changed over time, but the rate had formerly been slow enough for plants and animals to adapt over many generations; the changes now predicted would come about too swiftly for adaptation to take place. The 1971 report of the Study of Man's Impact on the Climate estimated that the annual global mean temperature would rise by some 0.5° C by the year 2000 and by some 2° C by about 2030, when the concentration of CO_2 would be twice its preindustrial level of about 280 parts per million (ppm). Later predictions, based on changing conditions and new scientific research, showed how difficult it is to predict future energy demands. The oil crises of the 1970s forced the industrialized nations to conserve fossil fuels, which led a 1980 conference at Villach, Austria, to estimate that in 2025 the concentration of atmospheric CO_2 would be about 450 ppm and that it would take almost 100 years to double the preindustrial level—a very different timeframe from the 1971 estimates at the height of oil usage. Then further research showed that other trace gases also contribute to global warming, and many scientists returned to approximately the 1971 figures.

In 1982 the UNEP, the WMO, and the International Council of Scientific Unions (ICSU) sponsored an investigation of the carbon dioxide problem whose results were discussed at a second meeting in Villach in 1985. The studies were unanimous in predicting that by 2030 the doubling of atmospheric CO_2 and other greenhouse gases would lead to a warming of an average of 3° C. Clearly, atmospheric CO_2 concentrations were climbing, mostly through industrial emissions. Emissions caused by deforestation and changes in land use contributed to the rate of increase but were not themselves major contributors to global warming. It was agreed that trace gases—in particular methane, nitrous oxides, other oxides of nitrogen, and chlorofluorocarbons—had similar effects to those of carbon dioxide, and that their levels were also increasing. Trace gases seemed to be playing a much larger role in the greenhouse effect than had been expected.

There was almost unanimous agreement that doubling the expected greenhouse effect would raise the global average surface temperature; discrepancies among the results of different modeling approaches were insignificant. And this was not simply an issue for the North. There was evidence that the developing countries would increasingly contribute to climatic warming and would certainly be affected by it.

It was clear to those attending the Villach meeting that any agenda for action must examine all the available options. Governments and industry needed to discuss the feasibility of reducing emissions of greenhouse gases and identify the industries producing them; the costs and benefits of various strategies needed to be weighed, especially of a radical shift away from fossil fuel consumption; and most importantly, there should be some agreement on what the socioeconomic impact of these strategies might be, how much impact was acceptable, and what options for response could be developed.

A strong commitment to further scientific and technical research was urged, because climate models and other projections needed to be greatly improved if they were to be a credible foundation for political action. What would be the effect of global warming on the average citizen in 20 or 30 years' time? The public must be kept informed.

In 1988, the WMO and the UNEP set up the Intergovernmental Panel on Climate Change (IPCC) to assess scientific information on the various components of climate change, to examine the environmental and socio-economic impacts of climate change, and to identify realistic policy responses. More than a thousand scientists from both the developed and the developing worlds participated, and their reports, published for the Second World Climate Conference in Geneva in November 1990, agreed that the climate is warming faster than at any time in the last 10,000 years. If the trend continues, average temperatures were projected to rise by an estimated 0.3° C each decade and about 3° C by the end of the twenty-first century, accompanied by a twenty-centimeter rise in sea levels by 2030, increasing to sixty-five centimeters by 2100. Low-lying coastal areas and coral islands would face catastrophe unless defended at incalculable cost; if only 1 percent of the six billion people expected to be alive in the year 2000 were affected, there would be sixty million environmental victims and refugees. Further effects would include changes in the distribution of pests and agricultural disease; disturbed weather patterns; increased forest deaths; and the impoverishment of biological diversity. The impact on human food supplies would be felt most severely in regions that are already under stress, mainly in developing countries.

Global warming and the forces driving it are now broadly understood. A clear scientific consensus has emerged on the range of warming to be expected during the twenty-first century, in spite of uncertainties about its precise regional distribution and its environmental consequences. Only two options are available, one to consider the issue academic and ignore it, which would lead to catastrophic consequences, and the other to take immediate action to slow the buildup of atmospheric greenhouse gases. In February 1989 a meeting of legal and policy experts was held in Canada to consider means of protecting the atmosphere. Canada advocated formulation of an umbrella convention on the protection of the atmosphere, whereas the UNEP favored a framework for action, giving priority to global warming, desegregating its causes, and identifying those that could be dealt with in the very near future, such as carbon dioxide emissions, chlorofluorocarbon production, and deforestation. The latter approach would, it was felt, buy time during which the world community could decide on measures to deal with more complex causes, such as methane and tropospheric ozone.

Following the IPCC meeting in 1988 at which the above understandings had been reached, a series of meetings was held over the next two years that culminated in the Second World Climate Conference. In the interim, UNEP Executive Director Mostafa Tolba was present at all the meetings, working intensely to convince the delegates of the urgency of the global warming threat. He advised establishment of a global framework to be broken down into individual protocols that would control the principal sources of greenhouse gases and deforestation—and in fact, the Global Convention on Climate Change, which was so intended, would be signed in Rio de Janeiro during the 1992 Earth Summit. Time and again, using all his powers of persuasion and rhetorical skills, he reminded his listeners of the complex nature of the undertaking as well as its urgency. Action taken now, he said, would by mitigating climate change help to address acid rain, deforestation, soil erosion, and ozone layer depletion. Programs in plant and animal breeding and in architectural design and regional planning would help speed the adaptive process; fuel prices should be made to bear their environmental and social costs, and fossil fuel use drastically cut. At a ministerial conference in Noordwijk, the Netherlands, in November 1989, he urged endorsement of a draft declaration that included reversing the deforestation that caused the loss of more than 13 million hectares of forest annually; implementation of the declarations provisions would result in a world net forest growth of 12 million hectares annually by the year 2000. The draft declaration also pointed out the need for technology transfer and a global partnership for sharing both costs and responsibilities for all the programs above. The sixty ministers present pledged to support afforestation efforts in both North and South.

At the World Conference on Climate Change in Cairo, December 1989, Tolba called again for control of greenhouse gas emissions, for technology transfer, and for reforestation. Referring to the problem of sea level changes as a result of global warming, he pointed out that coastal barriers, one possible response to rising sea levels, are extremely costly (the Netherlands spent $15 billion over 30 years and plans to spend another $10 billion by the year 2000) but need not be ruinous.

And he advised a multilateral mechanism immune to political manipulation to handle the vast sums needed to address world climate change.

A meeting of the IPCC in Washington in 1990 resulted in a declaration urging nations to take steps to reduce sources and increase sinks for greenhouse gases and—echoing the consensus of the UN General Assembly in 1989—suggesting a global convention on climate change. That there was a lack of strong motivation by this time among the ministerial delegations is indicated by the tone of the opening address by U.S. President George Bush, who called for additional research and avoided the use of the words "global warming."

Preparing the convention proved to be a formidable task. According to the World Energy Conference of 1989, global energy consumption will rise by 50 to 75 percent by 2020, using 1985 as a base year, of which the developing countries will account for up to 75 percent. This is due not only to population growth (China, Brazil, and India will have a combined population of 3.5 billion by 2050) but also to these nations' ambitious industrialization strategies, driven by fossil fuels. China, with one of the biggest coal reserves in the world, today contributes only 9 percent of global CO_2 emissions, but its development strategy will require a ninefold energy use increase by the year 2030. In the North, where a consensus is forming that market prices should reflect environmental and social costs, prices for coal could rise by $30 a ton and bring about a sharp reduction in its use. A U.S. EPA report in 1989 that reviewed such revised market prices predicted a rise of 20–40 percent for coal, 20–30 percent for oil, and 13–20 percent for natural gas. Alternatives to a reduction in coal use include removing carbon dioxide from stack gases at power stations (the estimated cost of a 50 percent reduction is $4 billion a year) and removing both sulfur dioxide and carbon dioxide (capital cost in the United States for a 50 percent reduction is estimated to approach $1 trillion). So it is not surprising that analysts prefer a shift from coal to oil to natural gas, in that order; in many scenarios, coal ideally disappears.

There is general agreement on the need to end the practice of forest burning and to increase reforestation efforts. Aside from the direct effects on atmospheric carbon dioxide levels of both approaches, forest conservation makes sound long-term economic sense.

In preparation for the Second World Climate Conference, it became clear that the challenge of dealing with climate change existed on three fronts: the industrialized nations, the developing nations, and the countries of eastern and central Europe. The industrialized nations needed to use the cost-effective technologies already existing to cut back on their emissions, especially the United States, which with about 25 percent of total carbon emissions is one of the least efficient energy users. Developing nations needed assurance that the developed world would provide the financial and technical means to help them avoid installing inefficient technologies. The countries of eastern and central Europe, burdened with outmoded, heavily polluting smokestack industries, needed help from the OECD nations in meeting the new targets, but it was important that such aid should not be at the expense of the needs of other developing nations. The cost of all these efforts was staggering, and both the scientific and diplomatic communities agreed that action would be less painful if they began with the development of better technologies and better uses for available ones. Some analysts argued that the initial steps could be economically beneficial, but longer term, more extensive measures might cost 1–2 percent of the global GDP.

At the November 1989 conference most of the industrialized countries had strongly supported concrete control measures for CO_2, with the United States disagreeing. By the meeting two years later in Geneva, four groups had formed: (a) oil-producing countries, led by Saudi Arabia, who would not hear discussions of CO_2 and refused to consider any possible reduction in the use of oil, joined by coal-rich countries and by the United States; (b) small island states likely to be swamped by a rise in sea level, who insisted on strong immediate action against all greenhouse gases and especially CO_2, since chlorofluorocarbons had been controlled under the Montreal Protocol, joined by developing countries with low-lying coastal areas; (c) other developing countries, primarily concerned with financing and technology transfer; (d) eastern and central European countries, which lacked the technologies for efficient energy production.

Informal consultations during the Noordwijk conference, the 1990 Geneva conference, and at Oslo in preparation for the UNCED failed to modify the U.S. position; likewise, the four groups held firmly to theirs while negotiations for the upcoming Earth Summit continued for fifteen months during 1991 and 1992. In the final analysis, the three major issues were the same as those that had occupied the Montreal Protocol negotiations: control measures, financial mechanisms, and technology transfer.

At this point in preparations for the Earth Summit meeting, for reasons that were never clearly stated, the convening governments removed the proceedings for the preparation of a Framework Convention on Climate Change from management by the UNEP. It has been speculated that the developed countries were not yet ready for the positive action and concrete measures so strongly advocated by the UNEP executive director; nevertheless he continued to press for firm commitments for reduction of greenhouse gases and for transfer of technology and financial resources and urged that the Convention on Climate Change not substitute words for action. In the matter of a basic choice between the objective of keeping temperature increases within a level that allows life on earth to continue or setting a maximum acceptable increase, the negotiating governments ultimately favored the second alternative. At the present writing, the three major goals have not been reached: reduction in the level of global emissions of CO_2 and greenhouse gases not controlled by the Montreal Protocol; establishment of energy efficiency targets and standards; flexible, innovative, equitable mechanisms for technology transfer and provision of financial resources to meet these standards.

The chairman of the Intergovernmental Negotiating Committee, Ambassador Raoul Estrada of Argentina, expressed the policy governing drafting the Framework Convention on Climate Change: "It would not be profitable to draft a treaty that lays out highly specific policies that only a few countries could agree to. Rather, they drafted a general treaty that sets an overall framework within which all governments can work together. The treaty offers governments a well defined process for agreeing, step by step, on specifications."

The Framework Convention was signed in Rio de Janeiro during the UN Conference on Environment and Development—the Earth Summit—in June 1992. It disappointed many environmentalists by its relatively conservative approach, which is to try to stabilize the concentration of greenhouse gases in the atmosphere at a level that will prevent dangerous changes in the climate system. The Convention stipulates that "[s]uch a level should be achieved within a time frame sufficient to allow ecosystems to adapt naturally to climate change, to ensure that food production is not threatened and to enable economic development to proceed in a sustainable manner."

In December 1997, five years after the Earth Summit conference, the contracting parties met in Kyoto, Japan, to adopt a protocol governing the emission of greenhouse gases. Scheduled to take place over ten days, the conference was forced to go into an additional day of negotiations before the various factions (the United States, the European Union, and Japan) could reach an agreement. Negotiations were marked by the same emotional and political crises that characterized earlier environmental conferences, and tensions were high until the very last day.

The protocol sets an overall target of a 5.2 percent reduction in emissions of six greenhouse gases by the period 2008–2012, using 1990 as a base year. It must be ratified by at least 55 countries, including all the developed nations, before it can enter into force. This compromise, although not completely satisfactory to any of the parties, is at least a first step. Under the agreement, however, some developed countries will be allowed to increase their emissions above the 1990 level, and it is unclear what methods will be used for joint implementation or trading in greenhouse gases, or how noncompliance will be determined and dealt with. Nor does the protocol cover how to handle the financial and technological needs of any developing countries that may commit themselves to it.

The real test lies just ahead: convincing the world community of the reality of the global warming crisis, clarifying the existing gaps and ambiguities in the new protocol, and achieving its ratification in time to ameliorate the most serious environmental problem facing our planet.

7

The Basel Convention on Hazardous Wastes

During the course of eighteen months, from October 1987 through March 1989, a series of intense negotiating sessions was able to achieve the first steps in the eventual establishment of an international protocol to regulate the transport and disposal of hazardous waste. These meetings highlighted the extent of disagreement on environmental issues between the developed and the developing nations and the determination of the latter to avoid future exploitation or control by the former.

The methods used to achieve even this degree of accord were basically private, informal meetings at which the delegates could examine the issues away from the glare of the political spotlight, and special pains taken by the chief negotiators to accommodate the concerns of the delegates who felt their countries to be most at risk. This chapter follows the course of the meetings that led to the Convention, but first it is necessary to lay out the nature of the problem as it existed in 1987 and as it continues, to a great extent, to exist.

The Issue

Many of the developing countries, particularly the "threshold" countries (fast industrializing developing countries), in their determination to overcome their economic difficulties and problems of population growth, have ignored the buildup of waste inevitably generated by industrial development, even as the industrialized nations struggled to manage their own waste streams. The sound disposal of hazardous waste is one of the most pressing problems on the international environmental agenda; well-publicized events such as the evacuation of residents near the Love Canal in the United States and the disappearance of contaminated waste from an industrial accident in Seveso, Italy, although they attract worldwide media coverage, are only isolated examples of the scope of the problem.

At the time of the Basel Convention, a globally accepted definition of the term "hazardous waste" did not exist. This compounded the problem of estimating how much hazardous waste was being generated and where: Statistics on waste generated in different countries were difficult to compare and often misleading. Though among the industrialized nations there was a wide exchange of technical and policy information on hazardous-waste management, very little information was available on the situation of waste disposal and hazardous-waste management in developing countries, where regional organizations in the field of waste management had only recently become active and still relied upon scattered and highly unreliable information, and where hazardous-waste management programs had yet to be developed. Most of the threshold countries had already experienced the results of careless industrialization and were thus more aware of the problems connected with negligent disposal of its by-products. Although a number of developing countries had established legislation and institutions to control industrial waste, including public health legislation, these had limited relevance to hazardous-waste management, where administrative and legislative powers were often nonexistent or available only on a regional or even local level. There was also the problem of financing hazardous-waste management systems, which was and is a very costly matter and was often considered an impediment to development.

Public and governmental concern over the problem was, however, growing. Many developing countries were in the process of establishing proper waste disposal practices, particularly with regard to municipal-waste disposal for congested areas. When thorough planning was undertaken for this purpose, the question of disposal of industrial and hazardous waste arose automatically.

Considerable advances have been made in hazardous-waste disposal technologies, including incineration and controlled disposal on land, but between one-half and three-quarters of the waste stream continues to be dumped on land. This proportion is expected to increase despite growing reuse and recycling of materials. Although landfill waste management has improved greatly, major obstacles remain. Thousands of landfill sites and surface impoundments have been found to be inadequate: corrosive acids, persistent organic chemicals, and toxic metals have accumulated in these sites for decades and are now leaking into groundwater and other media, posing serious health threats. In the United States, the Environmental Protection Agency identified 32,000 potentially hazardous sites, of which some 1,200 needed immediate remedial action. In Europe, 4,000 unsatisfactory sites have been identified in the Netherlands, 3,200 in Denmark, and a large number in Germany. Cleanup costs have been estimated at US $30 billion in the western part of Germany, $6 billion in the Netherlands, and about $100 billion in the United States.

Costs of new landfills are spiraling: in the United States it is estimated that landfill costs have increased tenfold in two decades. Local opposition to waste management facilities has become more vocal; existing disposal nightmares and ever increasing paperwork and costs have induced practitioners in some countries to send their hazardous waste overseas, and the international movement of hazardous waste has become big business. The majority of these shipments have been legal, but with the tightening of the controls over transport and disposal, illegal dumping and traffic have increased to become a global issue.

Most waste exports go from one industrialized country to another, with a consignment of hazardous waste crossing an OECD frontier on average every five minutes, for a total of more than 100,000 border crossings each year. In 1988, between 2 and 2.5 million metric tons of hazardous waste crossed European frontiers; annually some 200,000 to 300,000 metric tons move from western to eastern European countries.

Although at the time of the Basel Convention many industrialized countries had well-developed hazardous waste management systems and had enacted legislation to cover this issue, in general their import regulations were more comprehensive than their export rules. Some of these nations had established a special licensing program that would provide information on waste imports and permit refusal of such shipments. This left the developing countries vulnerable because of their lack of such protection and the unlikelihood that they would soon acquire environmentally sound facilities. A large number of unscrupulous—indeed criminal —waste brokers have exploited the price and regulation differences between developed and developing countries. The problem is particularly acute in Africa, where waste disposal costs at the most $40 per ton, whereas disposal in Europe costs between four and twenty-five times as much and in the United States twelve to thirty-six times that figure. Thus the world has witnessed the ugly—and immoral—spectacle of waste-laden ships sailing the seas in search of unsuspecting ports in the South, abandoning leaking drums of toxic waste at dockside in developing countries, or dumping it under cover of night.

It is generally agreed that in accordance with international law, every country has the sovereign right to prevent unwanted waste imports. Control, however, may be difficult without internationally agreed principles regarding notification and labelling, and controls are generally left to customs authorities, who may not be aware of specific problems and have no special powers with regard to hazardous waste.

By 1987, as the international waste traffic continued to increase, it had become clear that the problem demanded global solutions backed by rigorous national responses. In addition, the prospect of South-South traffic as developing countries pushed toward industrialization underlined the urgent need for prompt and effective action. Several journals of science and technology reported on the problem, and the media and NGOs took an active part in exposing the improper transfer of hazardous waste.

The varying levels of development of environmental law as well as the different disposal practices of different countries, highlighted by numerous incidents involving hazardous waste, called for the development of guidelines and common international measures for tracing and regulating the flows of hazardous waste between countries. This led to the negotiation of the Cairo Guidelines and Principles for the Environmentally Sound Management of Hazardous Wastes (see Chapter 4), adopted by the UNEP Governing Council in June, 1987.

While the world was absorbed in the negotiations of the 1987 Montreal Protocol, growing international concern, especially in the developing countries, led the UNEP Governing Council to adopt a proposal by Hungary and Switzerland that the executive director convene a working group of legal and technical experts who would prepare a global convention on the control of transboundary movements of hazardous wastes, drawing on the Cairo Guidelines, and to expedite the work of the group with the aim of finalizing the preparation of the convention as soon as possible. The council also requested the executive director to convene, in early 1989, a diplomatic conference to adopt and sign the resulting convention, a clear sign of the urgency governments assigned to dealing with the problem.

The organizational meeting was held in Budapest, Hungary, in October 1987, followed by six negotiating sessions in Geneva, Caracas, Luxembourg, and Basel from the beginning of February 1988 to March 1989.

The First Negotiating Session, Geneva, February 1988

Iwona Rummel-Bulska, who was active in the negotiations to establish the Cairo Guidelines, had presented a draft convention based on that document at the organizational meeting of the working group in Budapest and had revised the text for presentation at the Geneva meeting. She told the group,

> Executive Director Tolba indicated that in accordance with the Governing Council decision regarding the Working Group assignment, the draft Convention is not expected to be only a framework convention but that it should have direct practicable implications for the transboundary movement of hazardous wastes by specifying clearly the responsibilities of the country of export, country of import and transit country. The Executive Director is expecting the Working Group to give special consideration to the situation of developing countries and their need to receive technical assistance from developed countries in particular regarding developing countries' lack of capacity in assessing information received through notification about hazardous wastes and their lack of capacity for the safe handling and/or managing of hazardous wastes received.

At the February meeting the following points had been emphasized:

- the first revised draft convention prepared by the UNEP Secretariat was a useful starting point for the discussions.
- The convention should contain provisions for adequate notification procedures for importing and transit countries.
- The problem of the definition of hazardous wastes should be addressed as well as that of disposal.
- It should be determined whether wastes destined for recycling should be included in the convention.
- The industrialized countries have a special responsibility to help the developing countries implement the convention.

At its first session, the working group had elected a Swiss, Alain Clerc, as chairman, two vice chairmen, from Hungary (Jozsef Arvai) and Venezuela (Sibrahim Gerdler), and a rapporteur from Finland (Thomas Aarnio), all of whom kept their posts throughout the negotiations. All five regions of the world were represented at the first negotiating session: thirty-three countries, of which twelve were developing, and a number of delegations from international governmental and nongovernmental organizations.

The revised draft that emerged from the working group had some fifty words, phrases, paragraphs, or whole articles bracketed for later discussion.

The Second Negotiating Session, Caracas, June 1988

When the negotiating group met for its second session, forty governments attended, twenty-two of them from developing countries, an indication of the concern of the Third World over the issue of the transboundary movement of hazardous waste. During the review of the second revised draft convention, a number of issues emerged on which governments had diametrically opposing views, including the definitions of the terms "area under the national jurisdiction," "territory" and "generator"; whether radioactive waste should be included in the convention; and the impact of the convention on national legislation and on existing bilateral agreements.

Experts from member states of the Group of 77 present at the session requested the secretariat to include in the report the following five points, which were among their major concerns:

1. The interests of transit countries, both from the point of view of environmental protection and with regard to the health of their inhabitants, should be considered at the same level as those of the countries of import in the convention.
2. In order to ensure transparency in the transboundary movement of hazardous waste and to provide support to developing countries with a view to increasing their hazardous-waste management capacity, it is essential to have an effective and functional secretariat.
3. Although the role of the exporter and the producer has to be taken into account, the transboundary movement of hazardous waste must engage the responsibility of the country of export.

4. The territorial waters of countries concerned by the transboundary movement of hazardous wastes should be considered as an integral part of their territory.
5. Governments should cooperate closely to prevent clandestine and illegal transboundary movements of hazardous waste.

The observer from Greenpeace International expressed a concern that the convention as drafted could establish a mechanism for the export of hazardous waste from developed to developing countries and called for a worldwide ban on all exports of hazardous waste as the only guarantee for the protection of the global environment. This remained Greenpeace's position throughout the negotiations; lobbying vigorously for it, the organization convinced some African representatives, who later blocked the rest of the African states from signing the convention when it was finally adopted. Following the approval and signature of the treaty, Greenpeace continued its attack, bringing forward these same arguments during the final press conference.

The third revised draft that emerged from the second session contained fifty bracketed items, some of them unsolved from the second draft. It was obvious that progress on the convention was slow and that new issues would continue to come up as a result of the open and frank discussions that prevailed.

The Third Negotiating Session, Geneva, November 1988

Early in August 1988, in preparation for the third negotiating session, UNEP Executive Director Mostafa Tolba had circulated two notes to all governments giving concrete proposals regarding the convention. The notes included the following four points:

1. The UNEP proceeds from three premises: (a) that all governments have an interest in finding a clear solution to this problem that is creating friction and embarrassment among them, when the actions involved are mostly illegal movements of hazardous waste by small companies and individuals and not by governments themselves; (b) that hazardous waste should be disposed of close to the place of its generation in the country of origin and that transfrontier movements of hazardous waste should be allowed only under very strict conditions; and (c) that the UNEP was not concerned primarily with the transport of hazardous waste per se but

rather the end purpose of this transport, its disposal, which must be environmentally sound. (Up until that time the convention had dealt only with the transboundary movement of hazardous waste and had not touched on its disposal.)

2. The following elements should be considered for inclusion in the preamble to the convention: (a) the most effective way of protecting human and environmental health from hazards posed by toxic waste is the complete banning of the movement of such waste away from its origin; (b) hazardous waste should be disposed of close to the place of origin, and transboundary movements of such waste from the country of generation to any other country, in particular developing countries, whether or not contracting parties to this convention, should be kept to a minimum; and (c) such restrictions on movement of hazardous waste would act as an incentive for reducing the generation of such waste and also for disposing of it close to the place of generation.

3. The definition of hazardous waste produced by the working group at its meeting in Caracas is a big step forward. However, two points need to be taken into consideration: (a) developing countries rarely have standards or regulations to define what they consider hazardous waste; and (b) substances or materials that may not be considered hazardous in a developed country may be hazardous in a developing country, where the people are much less informed about, or trained in, the handling of waste.

4. In the body of the convention, under appropriate articles, the following points needed to be considered:

- The issue of export of hazardous waste from offshore territory.
- The shipment of hazardous waste under the flags of convenience or on board ships registered offshore.
- The obligation of the exporter to prove whether or not the waste possesses the hazardous waste characteristics listed in annex II to the draft convention, and the obligation of the government of the country from which the waste is exported in this respect.
- The role of the governments of export in ensuring that the notifications to countries concerned are prepared in conformity with the requirements of the convention.
- All practicable steps to be undertaken by the contracting parties to ensure that hazardous waste is disposed of close to its origin within the territory of the country where it was generated. When there are compelling reasons to dispose of it outside the country of origin, the parties should ensure that the transboundary movements and ultimate disposal of hazardous waste are conducted in a manner that will protect human health

and the environment against the adverse effects that may result from such movements and disposal.

• The assurance of the establishment of adequate treatment and disposal facilities by the countries in which hazardous waste is generated, in order to manage their hazardous waste close to the point of generation.

• The assurance that the transboundary movement of hazardous waste is permitted only under compelling conditions and that it should be kept to a minimum, regardless of economic benefits.

• The continuous reduction of the amount of hazardous waste exported to other countries, in particular developing countries, whether or not contracting parties to the convention, with the aim of completely eliminating the export of waste, and the monitoring and reporting by each contracting party to the secretariat of the convention on the total yearly amount of hazardous waste exported from its territory, as well as the provision of information to the secretariat about its efforts to achieve a significant reduction of the amount of hazardous waste exported.

• The further development of the format and nature of information to be included in the notification from exporter to the countries concerned, which format could be annexed to the convention.

• The further development of the details of the nature and modalities of implementation of assistance to be given to the countries concerned that need such assistance, in particular developing countries, in reviewing the notification received and in certifying that shipments of hazardous waste received by or transmitted through countries concerned, in particular developing countries, are identical to the material described in the corresponding notification.

• The further clarification of the modality of proving whether the country receiving the waste possesses adequate technical capacity for the disposal of the waste in question.

• The obligation for contracting parties to cooperate in the case of illegal transboundary movements, that is, movements not in conformity with the convention, ought to be addressed separately, as should consideration of the feasibility of setting harsh penalties and fines against those who practice illegal dumping of hazardous waste, of the role of the governments of the exporters in checking hazardous waste leaving their territories, and of the issue of the temptation of financial benefits involved both for business people dealing with hazardous waste and for the developing countries receiving it.

• The meaning of the following phrases needs to be clarified unambiguously: "accepted and recognized international rules," "standards or practices concerning environmental protection," "adequate proof," "transport safety aspects," and "tacit consent.

• There must be cooperation in the environmentally sound disposal of hazardous waste.

• Consultation on liability and compensation should refer to the damage resulting from transboundary movements of waste as well as its disposal.

• Should information on accidents also cover information on accidents during the disposal operation?

• There should be a stipulation that the convention be reviewed in the light of experience to be gained during its implementation, after a certain period of time after its entry into force.

• There should be clarification of the meaning of "a water body except seas/oceans," and whether this should include man-made freshwater lakes and marine lakes.

• Incentive is needed for countries to become parties to the convention.

• There must be unambiguous clarification that the convention does not cover radioactive waste.

• Procedures and institutional mechanisms should be devised for determining noncompliance with the provisions of the convention and treatment of parties found to be in noncompliance.

• Contradiction should be avoided in the article dealing with bilateral, multilateral, and regional agreements.

• Who will determine whether a hazardous waste does or does not possess any of the characteristics contained in annex II to the draft convention?

• Which country's legislation or consideration will determine whether the operations mentioned in the convention will be included or exempted from the definition of disposal?

Governments commented extensively on the two notes. Twenty-one governments—eleven of them from developing countries—responded with detailed comments on various proposals in the notes. The responses were collected and presented to the working group at its third meeting in November, together with the results of informal personal consultations with a small group of governmental experts and with high-level representatives of some preshipment inspection companies, some major industries, and a number of important NGOs. These informal consultations focused on the following ten issues:

- specific waste products to be covered by the convention
- issues of the responsibility of states, liability and compensation, and noncompliance with the provisions of the convention
- modalities of assistance offered to developing countries in checking notifications and testing shipments
- means of ensuring that receiving sites or facilities are environmentally sound
- action in cases of emergency
- illegal traffic in hazardous waste
- issues of offshore territories and ships carrying flags of convenience
- criteria for allowing transboundary movement of hazardous waste and for approving waste disposal fisites and facilities
- financial arrangements required for the implementation of the convention
- the issue of the lack of the required infrastructure, especially in developing countries

In these consultations the UNEP Secretariat, as represented by Tolba and Rummel-Bulska, once again took an active position, taking stands on the issues on behalf of the environment and of the poorer and weaker countries. New elements were introduced into the debate, particularly the issue of the disposal of hazardous waste, and active consideration began on the specifics in the negotiations.

The realization of the wide divergence of the views of countries and groups of countries showed the need for further informal consultations with individuals or small groups of government representatives. The negotiators agreed and began their own informal discussions. Here a group of young, active government representatives, who became friends over time during the negotiations, worked together to find compromise formulations. In fact the need to give the meetings an informal status required that some of them were chaired by Tolba or Rummel-Bulska rather than by the group chairman, Clerc. The young men and women among government representatives who played key roles in the negotiations included Bakare Kante (Senegal), Julio Bitelli (Brazil), Karen Jorgensen (Norway), Ahmad Fathallah (Egypt), and José Luis García (Spain).

The leader of the U.S. delegation, Andrew Sens, was extremely active and participated in a large number of informal consultations that helped solve a number of delicate issues. Two delegates, old hands in the negotiation although young in years, Patrick Szell (U.K.) and Michael Smith (Australia), played a very significant role in finding compromise formulations to some difficult subjects.

The report of the November meeting of the working group states:

> The Executive Director urged the group to work expeditiously to reach Basel on time for the signature of the Convention in March 1989. . . . He emphasized that the Working Group was not drafting a convention on the physical movements of hazardous wastes. . . . The aim of the Convention was to establish control measures that would:
>
> 1. lead to major reduction in the generation of hazardous wastes and thus eliminate the need for their movement; and
> 2. Make it very difficult to get approval of movement of hazardous wastes with the goal of reducing to a minimum their transboundary movement and of ensuring that such movement is only permitted when it is more environmentally sound to dispose of waste farther than close to where it is generated.

Fifty governments attended the third negotiating session in Geneva, fully thirty from developing countries. The time was getting very short before the date set for a plenipotentiary conference to adopt the convention, and governments were getting more interested and more concerned, while negotiations became much more difficult. Twelve very strong international NGOs, representing both environmentalists and industrialists, attended the meeting and were both active and vocal, each lobbying in a different direction. This made it necessary to hold informal consultations with each group: governments, shipping companies, and industrial and environmental NGOs.

Unavoidably, with the inclusion of new issues and the proposals above that had been supported by so many governments during the negotiations, the text of the fourth draft convention that resulted had 200 bracketed items instead of the previous fifty, a depressing development in view of the short period left until March 1989, the scheduled final meeting in Basel, barely four months away.

The Fourth Negotiating Session, Geneva, January 1989

The number and intensity of the informal consultations increased. Tolba and Rummel-Bulska circulated a note entitled "Points Identified by the

Executive Director for Further Consideration at the Informal Negotiations on Hazardous Wastes, 4–6 January, 1989, Geneva," proposing changes in the title of the Convention to include the words "management of waste" in the five preambular paragraphs and eighteen articles and suggesting an additional article. The proposals covered the issues of the responsibility of waste exporter and generator, offshore territories and the obligation of flag states and registration states, criteria for managing hazardous waste, the sovereign right of a state to ban the disposal of foreign hazardous waste in its territories, the duty to reimport, control of illegal traffic, protocols on liability and compensation, noncompliance, financial arrangements, technology transfer, and amendment of the convention and its protocols and annexes: most of the issues that had been bracketed in the fourth draft.

By that time it was clear that the African states, where most of the dumping incidents reported by the media had occurred, were determined to achieve a regional protocol banning the importation of hazardous waste into Africa, a collective action that was essentially a political move to ensure that no African country would independently accept hazardous waste. There arose serious concerns that this group of countries might opt to block an agreement on a convention. Flavio Cotti, then the Vice President of the Swiss Federation and Federal Counselor for Interior (with oversight on environmental matters), who had been the prime mover of the UNEP Governing Council's decision to start negotiations on the Convention and had nominated one of his own assistants, Clerc, as chairman of the negotiating group, stayed in close touch with developments. He had already offered Basel as host of the ministerial conference to adopt the convention, less than three months ahead at that time, and had convinced the Swiss government to set aside enough financial resources to enable developing countries to be strongly represented.

As time went by with seemingly little progress, Cotti, Tolba, and Rummel-Bulska became concerned. Late in 1988 they discussed the advisability of having a meeting proposed by Senegal that would focus on Africa to allow those delegates to air their concerns and to see if some middle ground could be reached, agreeing that whatever problems might arise in such a meeting, it was urgently needed. Barely two months before the date of the Basel Conference, in January 1989, Senegal convened the special meeting in Dakar. All African states were invited and nearly all of them attended at the ministerial level. A large number of environment ministers of the OECD countries also participated.

From the outset the meeting was tense, largely because of the actions of a small group of delegates who, far from observing the usual amenities, used confrontational tactics and angry, undiplomatic language to make their positions known. The OECD ministers, particularly Cotti, were taken aback but did not lose sight of the fact that they were all working for the benefit of the environment everywhere. This made it possible for the chairman, Minister Moctar Kebe from Senegal, Cotti, and Tolba to work out and get approval for a position statement that, although it required no commitment from the Africans, did not threaten to derail the negotiations. It was very difficult, very embarrassing, and very frustrating.

What the dissenting delegates demanded was that in place of a global convention, Africa make a regional arrangement that would not bind anyone but themselves. However, what the other delegates considered to be, at best, the unclear logic of a few African ministers prevailed, but rather than derail the effort entirely, it was agreed to let it stand.

Based on the results of the Dakar Conference, the Geneva January consultations and an earlier meeting (December 1988) of a technical subgroup, Tolba and Rummel-Bulska prepared a formal note to the working group for its fourth session in Luxembourg in January–February 1989 entitled, "Proposals by the Executive Director for Consideration by the Ad Hoc Working Group at Its Fifth Session." The proposals covered all the points presented in the note to the Geneva informal consultations with some additions, including the title, the preamble and twenty-six articles, as well as many of the bracketed articles, paragraphs, sentences, phrases, or words in the fourth draft.

The Fifth Negotiating Session, Luxembourg, January–February 1989

Representatives of fifty governments and a large number of intergovernmental and nongovernmental organizations attended the fifth negotiating session. The meeting was nearly disastrous. In spite of careful preparations and continued informal consultations during the meeting itself, only the preamble and twelve articles out of the then-thirty articles constituting the draft convention were discussed, and reservations were expressed by various governments on almost every one of them. The anxiety bordering on despair of some of the government representatives may be sensed in parts of the report that was issued at its end.[1]

This was the state of affairs on the February 3, 1989, exactly forty-five days from the opening date established for the Basel Conference, which was expected to adopt an agreed convention and open it for signature. The Swiss government, in particular Cotti, felt very uncomfortable; Tolba and Rummel-Bulska were near panic. The meeting ended close to midnight, and the two UNEP leaders left early the next morning, driving to the UNEP office in Geneva because the airports were closed by heavy fog. A small group of government representatives who took varying positions but were anxious to come to an agreement also drove to Geneva, and upon arrival, around 4 P.M., two days of nonstop informal consultations began during which the government experts were sounded out in their personal capacities on the proposed compromise formulations and asked for their suggestions. Another informal consultation was held in Geneva from March 8–10, one week before the last meeting of the working group.

The Sixth Negotiating Session, Basel, March 1989

When the working group met for the sixth negotiating session in Basel in March 1989 there were still three major difficulties: the position of Africa, or rather that of the small group of African governments who were leading resistance to the signature of the convention and even its adoption; the position of the United States, especially regarding the relation of municipal waste to hazardous waste, as well as the problem of national legislation and regulations that might be difficult to change if they contradicted the text of the convention; and a number of reservations on issues that needed final clarification by various countries, including definitions of areas under national jurisdiction, transit countries, illegal traffic, and so on.

On the basis of the earlier informal consultations, a "Note by the Executive Director on some Points of the Hazardous Wastes Convention Which Were Not Resolved at the Fifth Session of the Ad Hoc Working Group" was circulated, covering a number of very significant issues:

transit countries, offshore territories, flags of convenience, state responsibility, illegal traffic, and bilateral, multilateral, and regional agreements.

By the time the government representatives met in Basel on March 13, 1989, in addition to the organizational session there had been six negotiating sessions over a period of eighteen months and nine informal consultations during the last seven months or less. Ninety-six states had participated in one or more of the formal and informal negotiations, sixty-six of them developing countries, a very clear indication of their concern on the subject. Fourteen UN bodies, eight intergovernmental organizations, and twenty-four NGOs and industrial associations had participated. And yet the situation on March 13 was very delicate. In his opening remarks to the working group, Clerc presented on behalf of Tolba (who was prevented by illness from attending the first few meetings) several proposals based on the informal negotiations held in Geneva from March 8–10. Clerc reviewed the work that had been done toward agreed formulations, including extensive consultation with Tolba on a number of unresolved issues, for which the latter had proposed compromise formulations in a note to the governments. He further discussed the program of the meeting in the light of the approaching deadline and encouraged the group to strive toward its goal, an agreed final text of the convention to be submitted to the Diplomatic Conference.

Seventy-six governments were represented at the meeting of March 13. Compromise formulations or proposals for changes or deletions were accepted for all bracketed words, phrases, sentences, or articles. The two issues that had created difficulties for the United States were resolved by introducing the term "hazardous wastes and other wastes," the latter covering municipal wastes. The issue of national laws and regulations was included in the relevant articles. It was agreed that a number of other observations and queries were to appear as declarations by the concerned states at the time of the convention's adoption.

The Basel Convention, March 1989

The plenipotentiary conference convened in Basel on March 20 at the ministerial level, with Cotti as chairman, Tolba as secretary-general, and Rummel-Bulska as executive secretary. The problem of the African delegates' intransigence hung heavy in the air. The African ministers, several of whom had authorizations from their capitals not only to adopt the convention but also to sign it, held long caucuses, during which the half-dozen ministers opposed to the convention blocked members from signing by shouting and screaming.

The meeting at large began with a few amendments, additions, and deletions to the fourth draft proposed from the floor, which could probably have been dealt with in the meeting. Then came the position of Africa, presented by Morifing Kone, Minister of Environment of Mali, which was at that time at the head of the Organization of African Unity (OAU).

Kone reminded the conference that the presence of African delegations in Basel reflected their awareness of the gravity of the problem and the importance of addressing it. He emphasized the conviction of African nations that the dumping of toxic wastes in the African continent is a morally reprehensible and criminal act. Recalling efforts made by the OAU in recent years to address this problem, he mentioned in particular the discussions at the forty-eighth Ordinary Session of the Council of Ministers of Africa and the subsequent summit of its heads of state, which led to the adoption of a "Resolution That Condemns the Dumping of Nuclear and Industrial Wastes in Africa as a Crime against Africa and the African People." The resolution calls on African states to prohibit the import of such wastes and requests the secretary-general of the OAU to cooperate with the relevant international organizations to assist African countries in establishing appropriate control mechanisms. Kone also recalled the resolution adopted by the Council of Ministers of the OAU at its forty-ninth Ordinary Session, which called upon African states to adopt a common position in the negotiating process on the Basel Convention.

Although expressing his appreciation of the efforts of the international community to adopt a global legal instrument addressing the problem, Kone stated that African countries were not prepared to sign a convention at this stage. In particular, he expressed concern that because of the limited technical capabilities of developing countries, it would be difficult for them to use the Basel Convention to prevent unscrupulous individuals from engaging in illegal dumping activities, and that African countries could still be used as dumping grounds for foreign waste, despite the efforts of the OAU.

Kone and other African ministers then presented some twenty-four different amendments to the draft convention. The chairman, secretary general, and executive secretary agreed that it was imperative to resort to informal consultations with all ministers who proposed changes. This developed into a ten-hour marathon session, informal and open, in which the conference leadership was able to explain what parts of the proposals were actually included in the draft convention and how some of the proposed changes could be included. By the end of the day all issues were resolved, guaranteeing that the African states present would not object to the adoption of the convention.

On March 22, Cotti announced the results of the previous day's consultation to the conference, and the convention was adopted by consensus by the 116 states present; 105 of them and the European Community (EEC) signed the final act, including the decision to adopt the convention. After the declarations and the closing remarks, the convention was opened for signature. Thirty-five of the states present and the EEC signed the convention. To the disappointment of most delegates, the handful of African ministers in opposition prevailed: the other African ministers whose credentials carried the authorization to sign, abstained, seemingly in the name of solidarity.

The main provisions of the Basel Convention are as follows:

- Every country has the sovereign right to ban the import of hazardous waste.
- The control system provided by the Basel Convention ensures that no hazardous waste is shipped to a country that has banned its import.
- Exports of hazardous waste to nonparties and imports from nonparties are prohibited, unless subject to a bilateral, multilateral or regional agreement, the provisions of which are no less stringent than those of the Basel Convention.
- Every country has the obligation to reduce the generation of hazardous waste to a minimum and to dispose of it as close as possible to the source of generation. Transboundary movements of hazardous waste may take place only as an exception, if they present the most environmentally sound solution, and if they are carried out in accordance with the strict control system provided by the Convention.
- Transboundary movements of hazardous waste carried out in contravention of the provisions of the Convention are considered illegal traffic.

The Convention states that illegal traffic is a criminal act, and obliges states to introduce national legislation to prevent and punish it. A state responsible for an illegal movement has to ensure the environmentally sound disposal of the waste in question.

• Industrialized countries have an obligation to assist developing countries in technical matters related to the management of hazardous waste.

• The Convention also calls for exchange of information and international cooperation.

• It enshrines the principles of notification and prior informed consent (the importing country must advise the exporter in writing of its willingness to accept a shipment on the basis of detailed information of what it contains).

• It requires that the exporting country be assured that the shipment will be disposed of correctly, something that few national laws demanded until recently.

• Developed countries that signed the Convention agreed to provide technical assistance to developing countries, so that they could acquire facilities for handling and disposing of hazardous waste in an environmentally sound manner. The Convention Secretariat monitors and coordinates these activities.

The resolutions adopted by the Basel Conference requested further action in connection with enforcing and strengthening the provisions of the Convention, including cooperation with other organizations to harmonize the Basel Convention with other international legal instruments, development of elements for inclusion in a protocol on liability, and development of technical guidelines for the environmentally sound management of hazardous waste.

The Basel Convention is the only existing global legal instrument regulating transboundary movements of hazardous waste. Its provisions ensure protection of countries against uncontrolled dumping of toxic waste and promote environmentally sound waste disposal and minimization of waste generation. The control system ensures that the Convention does not remain a mere declaration of intentions, but that the rights of countries are respected.

The Basel Convention entered into force in May 1992 and represented a durable step toward more effective action for global environmental protection. Like the Montreal Protocol on Substances that Deplete the

Ozone Layer, it was designed to be strengthened in the future. This has always been the policy advocated by the authors in developing international treaties: to resort to framework conventions when they can be easily and quickly supplemented by protocols, or to develop conventions that contain most of the detailed actions needed, but certainly not to press for "all or nothing." We believe a reasonable goal was achieved: a flexible treaty that can be amended or adjusted in view of new facts or new information. As stated earlier, a bone of contention during the whole negotiation process was the issue of banning the transfer of hazardous waste from North to South. This was one of the reasons why all African countries refused to sign the Convention; they wanted first a collective ban on imports, and a ban on exports by developed countries.

In January 1991 the OAU adopted an African Regional Convention, the Bamako Convention, that bans the import of all forms of hazardous waste, including nuclear waste, into Africa and controls the transboundary movement of such waste generated in Africa.

Further Meetings

First Meeting of the Contracting Parties, Uruguay, 1992

The Parties to the Basel Convention met in December 1992 in Periapolis, Uruguay. The question of a ban on waste exports from developed countries was again raised very forcefully by developing countries and a number of NGOs, especially Greenpeace, which insisted on a complete ban. At that time most of the developed countries (the United States, Japan, and all EEC countries except France: the major producers of hazardous waste) were still in the process of treaty ratification and were not ready to consider the issue. It was obvious that without their becoming parties to the Convention, it would be worthless. With the assistance of a member of the Convention Secretariat, Ahmad Fathallah, Tolba resorted once again to the formula of informal consultations. For three days continuous separate meetings were held with developed and developing countries, conveying the views of one group to the other on the language proposed by the Secretariat team, while the main meeting went on under the chairmanship of the Conference President, the Vice Minister of Environment of Uruguay, Gulio C. Balino Cotelo, to consider other important issues such as a budget for the Convention Secretariat and a fund for assistance to developing countries.

Iwona Rummel-Bulska carried the responsibility of Executive Secretary of the Conference of the Parties, assisted by two members of the Secretariat, Asa Granados and Pierre Portas. Hard bargaining on the budget led the Conference to establish a working group headed by Mr. Lauri Tarasti, Vice President of the Conference, from Finland. The stalemate on the movement of waste from North to South caused a great deal of tension. In the last day of the meeting an agreement was reached on language that would separate the issue into the general movement of hazardous waste and the movement of hazardous waste destined for recovery operations. The issue of recyclables has been and still is extremely sensitive to a number of countries, some of whose industries rely to a large extent on the movement of such waste.

The Conference was then presented with two draft decisions, which were adopted. Decision I/16, entitled "Transboundary Movements of Hazardous Wastes Destined for Recovery Operations"

> requested its technical working group to review the issue and consider the views submitted by states and interested organizations, giving consideration to criteria that determine whether such wastes are suitable for recovery operations, to present its recommendations on guidelines, procedures, or other matters within the frame work of the Basel Convention to the Second Meeting of the Conference of the Parties for its consideration.

The second draft decision, Decision I/22, was also adopted by the parties. As originally presented, it read:

> • Recalling that the aim of the Basel Convention is to reduce to a minimum the generation of hazardous wastes and other wastes and ensure that whatever is produced is disposed of in an environmentally sound manner as close to the point of generation as possible,
> • Recalling also the fourth ACP/EEC Convention of 15 December 1989 Lome IV and the Bamako Convention on the Ban of the Import into Africa and the Control of Transboundary Movements of Hazardous Wastes Within Africa of 30 January 1991, which prohibit transboundary movements of hazardous wastes to developing countries,
> • Recalling further that the Lome Convention requires ACP States to prohibit the direct or indirect import of hazardous wastes into their territory from the Community or from any other country,
> • Conscious that, during the negotiations leading to the United Nations Conference on Environment and Development (UNCED), developing countries called

for the prohibition of hazardous waste shipments from industrialized to developing countries:

1. Requests the industrialized countries to prohibit all transboundary movements of hazardous wastes and other wastes for *disposal/final disposal* to developing countries;
2. Requests further industrialized countries to inform the Secretariat on measures undertaken in order to implement paragraph 1;
3. Requests developing countries to prohibit the import of hazardous wastes from industrialized countries;
4. Further requests the developing countries to inform the Secretariat on the measures undertaken in order to implement paragraph 3;
5. Requests the Secretariat to report to the second meeting of the Conference of the Parties on the information received pursuant to paragraphs 2 and 4 above.

What resulted from the marathon of informal consultations reveals how the concerns of developed countries were understood and finally accepted by developing countries: with changes underlined, the final text reads:

- Recalling the aims of the Basel Convention to reduce to a minimum the generation of hazardous wastes and to prevent the transboundary movement of such wastes if there is reason to believe that the wastes in question will not be managed in an environmentally sound manner,
- Recalling decision I/14 regarding the transboundary movements of hazardous wastes destined for recovery operations,
- Reaffirming the obligations of all parties, including industrialized countries, as provided for in the Convention, to prohibit the export of hazardous wastes and other wastes to parties which have prohibited their import and to non-parties,

1. Requests the industrialized countries without prejudice to paragraph 2 to prohibit all transboundary movements of hazardous wastes and other wastes for *disposal/final disposal* to developing countries;
2. Notes that until the Conference of parties receives and acts upon the report of the Technical Working Group referred to in Decision I/14 and until appropriate measures are taken pursuant to paragraph 7 of article 15, transboundary movements of hazardous and other wastes destined for recovery and recycling operations take place in accordance with the provisions of the convention and in particular the requirement that the waste be handled in an environmentally sound manner.

The issue of North-South movement of hazardous waste was so sensitive that a title for the draft decision could not be agreed on in the informal consultations, and it alone was left untitled among the other twenty-two decisions.

Second Meeting of the Contracting Parties, Geneva, March 1994

The issue of North-South movement of hazardous waste was revisited at the second meeting, barely fifteen months later, and again, out of twenty-seven decisions adopted at that session, it was the sole decision left untitled. This time the developed countries, a sizable number of which had meanwhile become parties to the Convention, reluctantly accepted the position of the developing countries and adopted Decision II/12:

1. To prohibit immediately all transboundary movements of hazardous wastes which are destined for final disposal from OECD to non-OECD states; and
2. Also to phase out by 31 December 1997, and prohibit as of that date, all transboundary movements of hazardous wastes which are destined for recycling or recovery operations from OECD to non-OECD states.

This affirmed the wisdom of the policy to seek a treaty that meets some of the ultimate goals or objectives of dealing with the specified environmental issue, provided that such a treaty has the flexibility to allow for its adjustment as information becomes more concrete, technological developments improved, and public pressure stronger. It worked well with both the Montreal Protocol and the Basel Convention.

Some General Remarks on the Negotiations of the Basel Convention

During the process of negotiating the Basel Convention it became clear, particularly at the later stages of the negotiations, that one of the most difficult tasks was to bridge the gap in confidence. Many developing countries believed that developed countries wanted only a legal act that would amount to lip service or window dressing and would have no real effect on the dumping of hazardous waste on developing countries' territories. This feeling was particularly clear whenever developed countries referred to the OECD draft convention that they tried to develop, or when some of the developed countries referred to their own laws as sacrosanct. The two main sources of conflict and sensitivity among developing countries were illegal traffic in hazardous waste and the liability and responsibility of the state of export. Ultimately both were resolved.

A second difficulty was that an environmental agreement cannot be reached in a vacuum. As had been found in earlier negotiations, particularly those for the Montreal Protocol, the environmental problem is relatively easy to identify, but once negotiators start to agree, concerns immediately arise on how to handle the agreement's potential effect on trade and economics. On the trade side was the issue of hazardous waste destined for recycling or recovery; on the economic side was the price to be paid, essentially by developed countries: Technical assistance must be provided; illegal traffic must be stopped and hazardous waste taken back by the states of export if it could not be disposed of in an environmentally sound manner; and the issue of liability and responsibility, including state responsibility, must be put on the negotiating table.

Another issue of marginal importance for this Convention, but a major issue to transit countries, suddenly emerged: the definition of the area under national jurisdiction as it is connected to the principle of innocent passage. The complexity of these issues could have destroyed the fragile agreement that was reached, and the problem was solved only at the the last minute, through intensive informal consultations.

At certain moments some delegations seemed to feel that Tolba was taking the side of developing countries, particularly on the issue of illegal traffic. In all such cases he adhered to what he considered to be the approach of the international organization under whose auspices the negotiations were carried out, and he remains convinced that if such support had not been forthcoming, several developing countries, in particular African countries, would have walked away from the negotiating table and the negotiations would have failed. An example of the emotional climate that accompanied some of the negotiations is the January 1989 meeting in Dakar, described above; in order to go as far as they did toward agreement, they needed to have confidence that they were receiving a fair hearing.

A Protocol on Liability and Compensation

A critical component of the Convention, particularly for countries that have been, and continue to be, victimized by illegal traffic, is the redress of damages incurred by transboundary hazardous waste shipments and disposal. In Basel, governments called for the early adoption of a liability protocol. The UN General Assembly attached particular importance to Resolution Three of the Basel Conference dealing with that subject, and requested that the executive director of the UNEP brief the summer 1990 meeting of the Preparatory Committee of the UN Conference on Environment and Development on the progress made on the issue. It further requested him to report on progress to the UNEP's Governing Council and to the next General Assembly.

But broad support for a liability protocol does not simplify the very complex issues involved. The working group established by the UNEP in 1990 to negotiate such a treaty was facing familiar challenges. Thirteen years before, the UNEP had established another working group to consider liability for transboundary pollution damage; similar work was undertaken by the UNEP to review liability and compensation related to offshore mining and drilling and natural resources shared by two or more states. More recently, the Montreal Ozone Protocol dealt indirectly with liability, through the question of noncompliance. None of these efforts went very far.

The UNEP has built up experience in this field, in which other international organizations, including the Council of Europe, the Helsinki Commission, and the OECD, continue to make progress. The International Atomic Energy Agency (IAEA) has been considering a joint protocol on liability under the Vienna and Paris Conventions for nuclear damage. The International Law Commission continues to make major contributions to international liability and compensation issues. In 1989, the ECE concluded a Protocol to the Convention on Civil Liability for Damages Caused During the Carriage of Dangerous Goods; the European Commission has elaborated a Directive on Civil Liability for damages caused by waste.

Of particular relevance was the IMO assessment of a separate liability convention, examining such critical questions as the desirability of a two-tier liability system, whereby public-sector finance would meet costs not covered by insurance and other markets; liability ceilings; the concept of loss of profit, regarded either as an extension of property damage or assessed under a separate category. To help focus discussion, questionnaires were sent out to governments on the existence, scope, and coverage of provisions for liability and compensation in bilateral treaties to which they were party, and the existence and nature of national legislation related to the transboundary movement and disposal of hazardous waste. Their responses, which were useful in putting the issue in focus, may be summarized as follows:

- Of all the responses, only three indicated the existence of bilateral agreements, of which two have liability clauses.
- Roughly half of the replies affirmed the existence of national provisions for liability for interterritory disposal regulations.
- The base for exoneration from liability is very diverse.
- The concept of damage is extended to the environment, or is unlimited, in four replies.
- In all replies except three, liability amounts are not capped.
- Time limits for claim submissions vary, from one year to no time limits.
- In two replies, compensation mechanisms included a compensation fund.
- The majority of replies showed recovery procedures based on definitions of pure economic loss for damages.
- Liability extensions beyond national jurisdictions were present with some limitations in five of the replies.

In another effort to facilitate the work of the negotiating group, a small representative group of legal experts was convened to advise on the elements to be included in a protocol. On the basis of this consultation, several areas seemed to require consideration, including channeling of liability; financial and time limits to and exoneration from liability; financial guarantees such as insurance; the need for a fund, its source and disbursement; claims; procedures; and finally the jurisdiction of domestic courts and applicable laws.

We do not know the exact amount of hazardous waste produced in the world today. But even conservative estimates indicate an annual and steadily growing production of these materials of more than 330 million tons. Until a protocol on liability and compensation is established and existing clean technologies and production methods are universally applied, industry will continue to produce increasing quantities of hazardous waste, and greedy, unscrupulous operators will continue to thrive.

The petrochemical and chemical industries of Europe and North America are currently responsible for nearly 70 percent of all hazardous-waste production, but worldwide production of chemicals is surging ahead. In 1950, the world produced 7 million tons of synthetic organic chemicals per year; today that figure is well over 250 million. Already, close to 100,000 chemicals are in common use, and each year sees a vast array of new formulations appear. All these activities result in more waste, and, unfortunately, more hazardous waste. No one wants these materials contaminating his own backyard; clearly, their transboundary shipment threatens to become an increasingly serious concern during the next decades. The process is well under way. Legislative controls in the industrialized countries, which appeared during the 1970s, were directed toward disposal, not recycling, and these combined with economic interests have had the perverse effect of causing a dramatic increase in the volume of traffic in waste.

The transport of such waste to developing countries is a cause for grave concern. As the cost of waste disposal in developed countries escalates, having now reached more than $2,500 per ton, in some African countries, where debt-ridden governments struggle to exist, exporters have been able to negotiate deals for "disposal" at a mere $2.50 per ton despite the lack of proper treatment facilities. We know the consequences of these activities. Reports, figures, and anecdotal evidence are readily available. But ominously, there is a veil of silence over the generation and disposal of hazardous waste. As an example of this apparent willful ignorance, a British politician remarked, "We do not know how much hazardous waste is produced . . . who produces it, what it is, and what happens to it." It is the same the world over, and it is a situation that cannot be allowed to persist.

Note

1. "The delegations of Finland, Sweden, France, the Federal Republic of Germany, and Norway, while emphasizing the importance of the global convention, expressed their concern over the number of crucial issues which have yet to be discussed.

"In order to prepare proper documents for full powers of signature for their ministers, they stressed the importance for them to know where they stand on matters of substance and to have a clear picture of the different positions on the remaining substantive issues. The delegations urged that the discussion on the remaining issues be kept focused on the point of substance.

"The expert from Lebanon stated that it is the understanding of his delegation that the statement made by Norway on behalf of several countries does not mean that a specific group or groups are hindering the proceedings of this working group. It is the responsibility of every participant in this exercise to bring the Convention through.

"The legal and technical experts of the Latin American countries present at the meeting stated that they are deeply concerned that no significant progress has been made in the discussions, owing basically to the fact that there is a tendency to ignore the consensus about the concept of territory already enshrined in the Cairo Guidelines for the Environmentally Sound Management of Hazardous Wastes, which served as the basis for these negotiations. This concept is of particular importance for determining the scope of the Convention.

"Consequently, in order to bring to a successful conclusion the process of preparing an international legal instrument which will be effective in controlling transboundary movements of hazardous wastes, they urge that the progress made in this area in Cairo be maintained."

8

The Loss of Biological Diversity

The earth's genes, species, and ecosystems are the product of more than 3,000 million years of evolution and are the basis for the survival of our own species. But the available evidence indicates that human activities are now leading to the loss of the planet's biological diversity and as a consequence are eroding the biological resources essential for future development. Given the projected growth in both human population and economic activity, the rate of loss of biodiversity is far more likely to increase than to stabilize.

In describing the treaty negotiations that resulted in the 1992 Biodiversity Convention, we shall use a number of specialized terms. For clarity, they may be defined as follows:

- *Ecosystems* are dynamic complexes of plant, animal, and microorganism communities and their nonliving environment, interacting as an ecological unit.
- *Biological diversity* refers to the variability among living organisms from all sources, including terrestrial, marine, and other aquatic ecosystems and the ecological complexes of which they are part—within and between species and among ecosystems. *Genetic diversity within species* refers to the variation of genes within species, as expressed, for example, in the thousands of traditional rice varieties in Asia. *Species diversity* refers to the variety of species within a region, measured either as the total number of species present (sometimes called *species richness*) or as a combination of species numbers and distinctiveness (*taxonomic diversity*).
- *Conservation of biological resources* means the preservation, maintenance, sustainable use, recovery, and enhancement of the components of biological diversity: *biological resources*, including genetic resources, organisms or their parts, populations, or any other biotic component of any ecosystem with actual or potential use or value to humanity.

• *In situ conservation* means the conservation of ecosystems and natural habitats and the maintenance and recovery of viable populations of species in their natural surroundings and, in the case of domesticated or cultivated species, in the surroundings in which they have developed their distinctive properties; *ex situ conservation* is the conservation of the components of biological diversity outside their natural habitats.

Conserving biological diversity is important for reasons of both principle and human self-interest. Human self-interest is involved because ecosystems function as the planetary life-support system, renewing atmospheric oxygen and playing a central part in the biochemical cycle. They are a source of food, fiber, timber, natural drugs, and other products; they conserve soil; and they shelter genetic strains to which crop breeders continually return in order to improve cultivated varieties.

The importance assigned to conservation of biodiversity depends on the values ascribed to species and ecosystems, hence on economic and political judgments, as much as on scientific understanding. The past two decades have seen a considerable growth in our understanding of the evolutionary processes that created the biological diversity on earth today and of the contemporary factors that are leading to its reduction. Trends in that reduction have been inferred and observed in many regions, and the importance of genetic conservation has been more and more widely accepted. The economic status of living natural resources has advanced greatly.

Over the past 3,000 million years, biological diversity has been steadily increasing, but this expansion did not proceed smoothly. The fossil record of the past 600 million years has shown several mass extinctions during which the general trend of growing diversity was rapidly reversed.

Throughout the Earth's history, major taxonomic groups have emerged and diversified, only to decline and be replaced by other groups. The causes of this growth in diversity and the periodic extinctions and replacements that occurred are still only partly understood, but it is generally agreed that it took place as continents drifted apart and rejoined, climates changed, new taxa evolved, massive volcanoes erupted, and asteroids struck the earth. The species that now dominate the earth belong to genera that appeared during the present (Cenozoic) era, which began about 65 million years ago. For most of the period, the appearance and extinction of genera have been roughly in balance.

Extinctions have always been a fact of life: Indeed, more than 99 percent of the species that have ever existed are now gone. But it seems clear that as humans became more numerous and moved into new areas, they caused a significant number of extinctions. Australia, the Americas, Madagascar, and New Zealand appear to have suffered much greater species losses in the past 100,000 years, as humans settled in those continents and islands, than has Africa, where our species evolved. The number of species now living on earth can only be estimated. About 1.4 million have been described, and of these some 750,000 are insects, 250,000 are plants, and only 41,000 are vertebrates. The best calculations suggest that the actual total is likely to be around ten million, but estimates range from two to one hundred million. Species extinctions have increased steadily since the sixteenth century, mostly due directly or indirectly to human activity. About 75 percent of the mammals and birds that have become extinct in the past 400 years were island-dwelling species, which are known to be especially vulnerable to introduced species and other new evolutionary pressures. Similarly, island floras tend to be far more fragile than continental ones; on several islands, more than 90 percent of the endemic plant species are rare, threatened, or extinct.

The Emergence of Biotechnology

For thousands of years, people have been manipulating the genetic wealth of biodiversity by selecting and breeding crops and livestock to meet their needs. The exploitation of plant and animal resources has been the mainstay of agriculture, forestry, and fisheries activities, from which a vast variety of domesticated animals and plants have emerged. But today, new biotechnologies are emerging that permit great increases in the efficiency of traditional breeding programs and that allow the modification of organisms in ways that were impossible using traditional techniques. Some of these new technologies, such as tissue culture, already have a record of application, but the most novel techniques, such as genetic engineering, are only today yielding their first commercial products.

The emergence of biotechnology and recent developments in recombinant DNA technology present a potential for links between conservation and sustainable utilization of genetic diversity. This relationship is one of mutual dependence. On the one hand, biotechnology has much to offer the conservation of biological diversity. It could lead to new and improved methods of preservation of plant and animal genetic resources and speed the evaluation of germ plasm collection for specific traits. On the other hand, maintenance of a wide array of biological diversity, and hence a wide genetic base, is important for the future of biotechnology and sustainable development. The genetic material contained in domesticated varieties of crop plants, trees, and animals and in their wild relatives is essential for breeding programs by which genes are incorporated into commercial lines for disease resistance and responsiveness to different soils and climates.

Biodiversity in the form of genetic variability in cultivated and domesticated species has become a significant socioeconomic resource. To cite just one example, new varieties of sugarcane in Hawaii have been required about every ten years to adapt to pests and to maintain productivity. Without the genetic variability that enables plant breeders to develop new varieties, global food production would be far smaller than it is at present.

Recent understanding of how genes are expressed in the plant and animal realms and of germ plasm biotechnology offer new possibilities for increasing the production of food, medicines, energy, specialty chemicals, and other materials and for improving environmental management. This reinforces rather than diminishes the need to maintain the richest possible pool of genes. As the field of biotechnology develops, the future needs for germ plasm will be far greater than now. The importance to the biotechnology industry of biodiversity conservation should not be underestimated. The projected loss of diversity could diminish the genetic base required for the continued improvement and maintenance of currently utilized species and deprive us of the potential to develop new ones.

The issues relating to access to biological diversity and biotechnology are important and complex. The potential of genetic diversity can best be exploited when genes are accessible to all users and when the information and technology on how to use them is equally transferable to all. Having a particularly valuable genetic resource or the technical capabilities to develop new varieties must not imply exclusive rights of ownership, although free access does not imply free of charge. Two-thirds of all species are found in developing countries, particularly in the tropics, but it is the developed nations that have most of the biotechnological tools needed to exploit them. Whereas possession and custody of a potential genetic resource might be limited to one nation, benefits can accrue to all nations. Accordingly, a fair balance of benefits between owner and consumer must be attained, involving cooperation, with reciprocal benefits between developing countries and industrialized countries. The new biotechnologies, by allowing increased use of the genetic diversity of wild and domesticated species, increase the value of the world's biodiversity. In recent years, biotechnology has been seen as the direct channel by which developing countries can, in a practical way, tap their enormous biodiversity for economic development.

For example, although wild species and the genetic variations within them make contributions to agriculture, medicine, and industry worth many billions of dollars per year, biodiversity tends to be perceived largely in scientific and conservationist terms rather than in economic and resource terms. This situation is partly the result of the fact that biodiversity researchers and managers have often been unable to provide sufficient evidence of the socioeconomic benefits of nature conservation. Careful analysis of its full costs and benefits often justifies much greater investments in conservation.

Some resources are consumed directly, without passing through a market; these—called *consumptive use values*—are often the foundation of community welfare in rural areas. For example, firewood and dung provide more than 90 percent of the total primary energy needs in Nepal, Tanzania, and Malawi and exceed 80 percent in many other countries. One study of four indigenous Amazonian Indian groups found that they used from one-half to two-thirds of all forest trees as food, construction material, raw material for other technology, and medical and trade goods. (Virtually all species were used as firewood or as food for harvested animals.) In Africa, harvested species help feed rural people, especially the poorest villagers living in the most remote areas. In Botswana, more than 50 species of wild animals provide animal protein exceeding 90 kilograms per person per year in some areas. In Ghana, about 75 percent of the population depends largely on traditional sources of protein, mainly wildlife, including fish, caterpillars, and snails. In Nigeria, game constitutes about

20 percent of the mean annual consumption of animal protein by people in rural areas. Wild sources account for 75 percent of the animal protein consumed in Zaire. Conventional measures of economic performance, such as GNP, have tended to ignore this very extensive use when calculating the annual income of such groups, even though the cost of replacing such goods from other sources would be considerable.

On the other hand, the term *productive use value* is assigned to products that are harvested commercially for exchange in formal markets. This is often the only value of biological resources reflected in national income accounts. Productive use of such biological products as fuelwood, timber, fish, animal skins, musk, ivory, medicinal plants, honey, beeswax, fibers, gums, resins, rattan, construction materials, ornamentals, animals sold as game meat, fodder, mushrooms, fruits, dyes, and so forth can be a major component of national economies. Such values can be remarkably high. Some 4.5 percent of GDP in the United States is attributable to the harvest of wild species, amounting to some $87 billion per year from 1976 to 1980. The percentage of contribution of wild species and ecosystems to the economies of developing agrarian countries is usually far greater than it is for an industrialized country. Timber from wild forests, for example, is, after petroleum, the leading foreign exchange earner for Indonesia. And throughout the tropics, governments have based their economies on the harvest of wild trees. The total exports of wood products from Asia, Africa, and South America averaged $8.1 billion per year between 1981 and 1983.

Although market prices represented by productive-use value can be an important indicator, the market price is not always an accurate representation of the true economic value of the resource, and it does not deal effectively with questions of distribution and equity. It is also apparent that consumers may value resources differently: for example, tropical forests are valued very differently by consumers of scenic beauty than consumers of lumber products. The methodology for defining and relating these different valuations is still being developed. Species with neither consumptive nor productive use may nevertheless play important roles in an ecosystem, supporting species that do have such uses. In Sabah, Malaysia, for example, recent studies suggest that high densities of wild birds in commercial albizia plantations limit the abundance of caterpillars that would otherwise defoliate the trees; the birds require natural forests for nesting.

All species are parts of ecosystems that provide services of very considerable but seldom calculated value to humans. These services are often public goods that benefit the entire community or the whole world without a cost being assessed.

Moreover, as has been demonstrated in Nepal, although these benefits may be enjoyed within the country itself, many benefits from conservation are realized outside the country's borders, in forms as diverse as reduced flooding because of the protection of upland forests, the supply of medicinal plants and genetic material, or the pleasure given to international tourists. To date there has been no concerted effort on the part of economists and biologists to establish an international program to evaluate these aspects of the value of biological resources, although such a program has been discussed.

Information is gradually accumulating on the economic benefits to be derived from using genetic diversity, showing the relevance of an economic evaluation of the added value of biological resources. They include:

• *In agriculture*: In Asia, by the mid-1970s, improvements using genetics had increased wheat production by $2 billion and rice production by $1.5 billion per year by incorporating dwarfism into both crops. A Turkish wild wheat plant with no commercial potential was used to give disease resistance to commercial wheat varieties worth $50 million annually to the United States alone. A single gene from an individual Ethiopian barley plant now protects California's $160 million annual barley crop from yellow dwarf virus. Major cultivators of crops improved by wild genes have a combined farm sales import value of $6 billion a year in the United States. An ancient wild maize strain from Mexico, a perennial resistant to seven major corn diseases that can also grow at high elevations in marginal soils, can be crossed with modern annual corn varieties, with potential savings to farmers estimated at $4.4 billion annually worldwide.

• *In medicine*: Of all useful plant-derived drugs, only ten are synthesized in the laboratory; the rest are still extracted from plants. In the industrialized world, the retail sale of plant-derived drugs was estimated at $43 billion in 1985, and it is estimated that the Western market for herbal drugs could reach $47 billion by the year 2000. In 1960, a child suffering from leukemia had only one chance in five of survival. Now, the child has four chances in five, due to treatment with drugs containing active substances discovered in the rosy periwinkle *Catharanthus roseus*, a tropical forest plant originating in Madagascar. Commercial sales of drugs from this plant now total around $100 million per year worldwide.

Species richness generally increases in magnitude as we move from the poles to the equator. In one fifteen-hectare area of the Borneo rain forest, for example, approximately 700 species of trees have been identified, equivalent to the total number of tree species in North America. Yet tropical forests home to roughly one-half of our planet's entire biodiversity inventory are being dismembered by as much as seventeen million hectares per year.

Tropical forests are not the only rich ecosystems. Wetlands, the Mediterranean climate regions in southern Africa, coral reefs, and temperate forest zones are likewise both rich in biodiversity and under severe ecological stress. Eighty percent of the 23,000 species of plants estimated to occur in South Africa, Lesotho, Swaziland, Namibia, and Botswana, countries with a Mediterranean climate, are endemic to the region, giving the area the highest species richness in the world, 1.7 times greater than that of Brazil. Wetlands provide essential breeding habitats for many species of flora and fauna and help to regulate water flows. And yet more than half of all U.S. coastal and freshwater wetlands have been destroyed. Many parts of Europe have lost nearly all their wetlands.

Human activity is having catastrophic impacts on biotas, habitats, and entire ecosystems. To destroy a unique habitat means to sentence to death all species that rely on that habitat for survival. But extinction is not simply a matter of overt habitat destruction. Covert, insidious destruction through air and water pollution, acid rain, spiraling toxic wastes, urban expansion, and demographic momentum are reducing the critical margin necessary for the survival of many species.

Overexploitation and the introduction of nonnative species into ecosystems are also important causes of extinction. The expected climate change and global warming could have catastrophic impacts on our planet's biological diversity. A report submitted to the Intergovernmental Panel on Climate Change suggests that global warming could lead to a decrease in net forested areas, as long-standing boreal and other forests face increased mortality and fire frequency as well as increased soil erosion and nutrient losses.

A study by the World Wide Fund for Nature (WWF) suggests that rising sea levels due to climate change could threaten almost all existing wetlands in the world, destroying nearly 80 percent of U.S. wetlands by the twenty-second century. In Asia, more than half a million hectares of wetlands are lost each year because of damming, pollution, and irrigation. Similar though less severe trends are underway in Europe and in North and South America.

Wetlands destruction is one of many complex threats to biological diversity. Another is current demographic trends. Today's population of more than 5.6 billion is expected to grow to pass the 6 billion mark before the year 2000. The most noteworthy impact of this source of destruction is the startling disappearance of rain forests. Current trends suggest that by the end of this century, anywhere between one-third to one-half of all remaining tropical forests could be so disturbed as to be unable to support their wealth of species.

James Lovelock, originator of the Gaia Hypothesis, summed up the prospects of a genetically sparse planet very well. You cannot have a sparse planet, he said, any more than you can have half an animal. Fifty years ago Aldo Leopold, father of the modern environmental movement, explained the need for diversity in a different way. He suggested that there was a land community, of which certain members were of economic use to humans while other members were not. But regardless of how humans valued its constituent parts, he argued, the land community must function as a whole. If man failed to see that, then humans would pay the price.

Tropical forests, coral reefs, and abyssal ocean plains teem with diverse life; half of the planet's species live in tropical forests, an area that constitutes only 6 percent of the Earth's total land surface. To take appropriate action to correct our destructive habits, we must know something about conservation of biological resources, global conservation needs and costs, and the economic forces that influence them.

The main defense for biological diversity remains the approximately 4,500 areas of in situ conservation areas, reserves that cover 485 million hectares worldwide but still represent less than 3 percent of the world's total ice-free area. Not enough in themselves, many of these reserves constitute little more than lines on a map, underfunded and understaffed. They were established to meet an adequate target of parks, reserves, protected landscape, and botanical gardens while taking into account the needs of local people and providing fair compensation to those who contribute these areas to the benefit of mankind. Moreover, although such measures of conservation are urgently needed, in situ policies in themselves cannot quarantine biological diversity against airborne pollution, ozone layer depletion, coastal and marine pollution, soil degradation, climate change and other environmental problems. In situ and ex situ conservation strategies that map out concrete goals, targets, and timetables should go together. We need to ensure that sizable areas of virgin forests, soils, coastal areas, oceans, and other natural resources are conserved, and that threatened species are protected, either in situ or in accessible gene banks. We also need to coordinate in situ and ex situ conservation plans and targets with ongoing or proposed action in related areas that affect habitats, ecosystems, and biological diversity: plans to halt desertification, conserve tropical forests, reduce transboundary air pollution, and protect freshwater resources, to name just a few.

Biological diversity has always been viewed as a common heritage, like knowledge, in which increased consumption by the few was assumed not to reduce its availability to the others. Such assumptions now appear inadequate. To correct them, clearly delineated economic incentives related to species conservation must be introduced, including a clarification of global conservation needs and costs. The stock of genetic resources represents a vast, untapped reservoir that can improve agricultural practices and provide disease-curing pharmaceutical and natural products. Today humans use less than one-tenth of one percent of all naturally occurring species. For species with high market values, investment in conservation is sound economics. The potential market value of millions of undiscovered species of value to the pharmaceutical, biotechnology, tourist, and other industries, though great, is undetermined; its promise must prompt us to increased levels of collective investments in conservation, even though this requires attaching a value to both known and undiscovered species. Some argue it is immoral to put a price on nature's creations, and this may be partly true. But we must learn to adjust economic systems to work for, not against, the functioning of natural systems. With the careful development of environmental cost-benefit analysis, intangible benefits such as intergenerational benefits of habitat and species conservation will not be undervalued.

The full penetration of the cash economy into the rural economy is still relatively new in many parts of the world. Although that penetration is not altogether a bad thing, one of its effects has been to create a vastly expanded cash market in seeds. For 12,000 years—from when the Sumerians first began to sow crops, almost up to the present day—farmers have been plant breeders, saving their best seeds for use the following year. Now the number of plant breeders has fallen from millions to a small cadre of scientists in the employ of a handful of transnational corporations, and seeds are a $93 billion per year corporate enterprise. One of the instruments that facilitated corporate penetration of the seed market is the legal regime of plant breeder rights. These laws vary from country to country, but they are normally an outgrowth of intellectual property rights, giving corporations legally protected property rights over particular varieties. These legal protections are seen by some as vital for stimulating the burgeoning growth of existing biotechnologies such as tissue culture and new techniques such as genetic engineering. It would be ironic if this system of property rights were to be extended to Third World countries, where most seed gathering takes place. There farmers would find themselves paying dearly for crop varieties bred on the basis of genetic material they themselves had provided. Although intellectual property rights can be an important spur to innovation, a market-based trade in genetic material should be a fair trade: one in which there is remuneration not only for the innovator at the very end of the process but also for the conserver who has preserved and improved the material over generations and now makes it available. The present market system places a high price on the final, value-added product but almost no value on its genetic precursors, which have often been nurtured and improved for generations in the country of origin. More than 90 percent of the wild seed varieties gathered in the last ten years have been from the Third World, but almost all of the benefits of those resources have gone to developed countries, where many are very rich, and almost none to the people of the donor countries, where most are very poor.

Some of the causes of the decline in biodiversity have already been addressed. The number of protected areas in the world nearly doubled since the 1970s, and the total land area under protection has increased by more than 60 percent in the same period. Genetic conservation has also been addressed. The International Board for Plant Genetic Resources (IBPGR) was set up in 1984 and has played a role in developing strategies for conserving crop genetic resources and the establishment of seed banks. The FAO has also been active in this field, and the emergence of farmers' rights as a legal counterpart to breeders' rights has largely grown out of their initiatives. Economic support for the conservation of biodiversity has also been developing. Debt-for-nature trades have been used in Costa Rica, Ecuador, Bolivia, and the Philippines. The Global Environment Facility set up by the World Bank, in cooperation with UNEP and UNDP, includes several hundred million dollars set aside for the conservation of biodiversity. A number of legal instruments have been established. The Convention on Migratory Species provides a certain amount of protection to some very vulnerable members of the land community. And CITES provides a safety net for some of the world's most endangered species, including most recently maximum protection for the African elephant.

The Technical Working Group on Biodiversity, 1988–1990

The UNEP Governing Council, convinced that there was a need for a global convention to preserve biodiversity, in 1987 established a working group to investigate the desirability and possible form of an umbrella convention to harmonize current activities in the field of biodiversity and to address other areas that might fall under such a convention, filling gaps in current laws. With Veit Koester (Denmark) as chairman, the group met four times: in November 1988 and February, July, and November 1990. UNEP Executive Director Mostafa Tolba coordinated the meetings, as well as those to negotiate the Convention, and Iwona Rummel-Bulska acted as executive secretary throughout, until the signing of the Convention on Biodiversity in Rio de Janeiro in June 1992.

In August 1988, in advance of the first meeting of the working group, Tolba convened a meeting of internationally known scientists in the field of biological diversity and sought their views on the subject. Chaired by Martin W. Holdgate, the group included Thomas Lovejoy, Jeffrey A.

McNeely, Kenton R. Miller, David A. Munro, Perez Olindo, Marinella Paes de Carvalho, Peter Raven, and Michael E. Soule. Additional UNEP participants in the meetings of the senior scientists were Genady Glubev, Ruben Olembo, Hamdallah Zedan, Bernardo Zentilli, and Mona Bjorkland. The group exchanged views on the meaning of biological diversity and the need to conserve it, actions to be taken and their priorities, the scope of existing conventions, and possible features of a global convention. They concluded that a global convention would be a powerful catalyst drawing together the efforts of the various sectoral and regional conventions in this field by giving overall shape and strategic direction to the world effort but agreed that such a global convention must not be adopted as a substitute for action, or it would blunt and deflect the efforts the world needs. Accordingly, they urged that any convention should be designed to

- have a sound basis in science;
- be truly comprehensive in scope, covering in situ and ex situ conservation and the protection of the biosphere from all significant damaging impacts, in harmony with and supplementing existing conventions in this field;
- be practical in defining obligations and goals, leaving the contracting parties the responsibility of achieving them;
- have the commitment of governments to funding at a realistic level;
- provide realistically for the transfer of resources, allowing its implementation by the poorer countries that are also the custodians of much of the biological heritage of the earth; and
- be capable of catalyzing and coordinating the efforts of governments and other agencies under other conventions in this field.

The report of the senior scientists, including some possible elements of a legal instrument, served as the basis for the ensuing discussions and recommendations of the working group. Their conclusions covered only the conservation of biodiversity and made no mention of such issues as access to biological resources or biotechnology, nor of sharing of profits.

The Technical Working Group: November 1988, February and July 1990

The first meeting of the working group of experts in biological diversity took place in November 1988 and was attended by experts nominated by the governments of twenty-five countries, eleven developed and fourteen developing, from every region in the world. It was also attended by representatives of a number of intergovernmental and nongovernmental organizations, essentially UNESCO, the FAO, the secretariats of CITES, RAMSAR (the Wetlands Convention) and the Convention on Migratory Species (CMS), as well as the IUCN (International Union for the Conservation of Nature) and the WWF.

This would be the last convention whose negotiations were sponsored by the UNEP with Tolba as its executive director. In his opening statement he outlined five main areas for consideration: the scope of conservation, increased scientific research, economic values, financing and technology transfer to ensure protection of genetic diversity, and access to genetic resources and to relevant technologies. He drew the attention of the meeting to the issue of access to biological resources, saying that the group would have to consider how to use both FAO plant breeders' rights and farmers' rights to promote conservation of biological resources, especially in the global South, as well as the definition of preferential criteria for access by owners of genetic resources to gene banks and to biotechnologically manipulated resources.

At that meeting the working group concluded that existing conservation conventions and other relevant programs were sectoral and did not cover the full range of biological diversity, and that the amendment of existing conventions for the purpose of achieving rationalization or consolidation of resources and for adequately meeting the full range of biodiversity at a global level would be extremely difficult and time-consuming, because the political contexts of the currently existing legal instruments differed, with different adhering parties, clients, and administrative provisions. Amendment of the existing instruments being neither possible nor desirable, there was an urgent need for a new international legal instrument and other measures for the conservation of biological diversity.

Following this meeting the UNEP Governing Council reemphasized the need to conserve biological diversity on Earth by the implementation of existing legal instruments and agreements in a coordinated and effective way and the adoption of a further appropriate international legal instrument, possibly in the form of a framework convention. The Governing Council also requested the executive director to convene additional working sessions of the group to consider the technical content within the broad socioeconomic context of a suitable new international legal instrument and other measures that might be adopted for the conservation of the biological diversity of the planet. The Council further requested the executive director to expedite the work of the group with the aim of having the proposed new international legal instrument ready for adoption as soon as possible. There still had been no reference to provisions for access to biological resources or to technology, including biotechnology.

The second meeting of the working group, in February 1990, was attended by forty-one countries, twenty-three of them developing, a more than 60 percent increase over the first meeting. Where the first meeting had had two vice presidents and a rapporteur, the second elected four vice presidents and a rapporteur, with one member of the bureau from each of the five regions of the world, the first indication that governments were concerned to have balanced representation. This composition continued over its third session. Opening the meeting, Tolba reiterated the need for a global effort in which developed and developing countries infuse a new spirit of cooperation into the North-South dialogue for the conservation of biological diversity as a fundamental element of environmentally sound and sustainable development. He outlined basic issues to which attention must be given in order to develop recommendations on how to deal with them in the proposed new international legal instrument on biological diversity: the nature of the international legal instrument, global conservation needs and costs, financing mechanisms, preferential treatment for those having control over genetic resources with respect to gene banks containing them and to essential newly developed varieties obtained through breeding them, and international transfer and favorable access to biotechnology that could be usefully applied or adapted to developing countries' needs. This last issue was the first reference made in these meetings to technology and biotechnology. Tolba stressed that any new international agreement should not infringe on the sovereignty of nation states over their natural resources but must protect the interests of the states in which the resources are located and provide incentives for conservation of biological diversity without inhibiting growth or sustainable development.

The second meeting of the working group made significant progress on a number of basic issues, discussing all the above points and identifying areas of basic conservation and utilization needs, as well as the need for and scope of financing that would lead to measures for implementation and funding through the adoption of a new legal instrument on biological diversity. The group concluded that the instrument should aim to incorporate concrete and action-oriented measures for the conservation and sustainable use of biological diversity, and requested the UNEP executive director to commission several studies as a means of responding to specific issues in the process of developing the new legal instrument. These studies covered global biodiversity conservation needs and costs; current multilateral, bilateral, and national financial support for conservation of biological diversity; an analysis of possible financial mechanisms; access to genetic resources and biotechnology; and biotechnology issues.

The results of the studies were presented to the working group at its third session, in July 1990. The goal was to consider negotiation issues in sufficient detail to begin drafting the legal instrument. Ruben Olembo, at that time deputy assistant executive director of the UNEP, read a statement on behalf of the executive director that stressed the need for gene-rich developing countries to work in tandem with technology-rich developed countries as the basis for an arrangement that would benefit both North and South, who would receive mutual benefits from cooperation for the conservation and sustainable utilization of the planet's biological diversity. The statement outlined five main areas that needed to be reviewed in order to establish their technical feasibility: conservation costs; financial modalities; technology transfer, especially biotechnology transfer; draft elements for the proposed convention; and the relationship between the proposed convention and existing global and regional conventions, agreements, and action plans on biological diversity.

This third meeting was attended by 78 countries, more than three times the number that attended the first meeting and almost double the number at the second, a clear indication of the interest governments were beginning to give to the subject. The executive director's note to the third meeting called attention to a number of issues, first pointing out that a new international legal instrument on biological diversity, in the form of a framework convention, should be comprehensive in scope, covering the full range of biological diversity at the intraspecies, interspecies, and ecosystem levels and addressing both in situ and ex situ conservation and protection of biological diversity from all significant damaging impacts; it should be in harmony with, coordinate, catalyze, and supplement the efforts of governments and other agencies under existing agreements in this field; it should contain as much technical, financial, and administrative information as possible, with a commitment for implementation.

It was evident, he said, that a political commitment to identify a specific financial mechanism or fund was essential for the success of the planned legal instrument and the global cooperation needed to conserve the biological diversity of the planet. Such a mechanism or fund should realistically provide for a transfer of resources to enable the poorer countries to abide by the convention. These nations are the owners and custodians of most of the biological resources constituting the biodiversity of the Earth, and the lack of complete information on the likely total cost of meeting the needs of global conservation of biological diversity should not delay the creation of a financial mechanism or become an obstacle to reaching a decision on such a mechanism. The convention should stipulate that such a mechanism must operate under the authority of the contracting parties, combining funding and clearinghouse functions, dealing with the design and implementation of biological diversity conservation activities and facilitating the transfer and development of relevant technologies. The working group was invited to consider different options on how to administer such a mechanism or fund and which organization to entrust with the lead role.

The implementation of programs approved by the parties should be a cooperative responsibility of a number of bodies, for example the FAO, UNESCO, UNEP, IUCN, and WWF. The contracting parties would assign responsibilities and allocate funds among the participating organizations, and the secretariat of the convention would facilitate cooperation between the organizations and prepare proposals for submission to the parties. The convention must also identify mechanisms to permit access to genetic resources and relevant biotechnology techniques, processes, and products while protecting the sovereign rights of states concerning their natural resources and the legitimate interests of biotechnology inventors.

It was essential, Tolba continued, that the planned legal instrument provide a link with existing conventions, agreements, and action plans relating to the conservation of biological diversity, benefitting from the experience of existing bodies. To this end, there should be an exchange of information and documentation; standardization of formats for reporting to and from contracting parties; regular preparation of an overview report of activities carried out under existing conventions, agreements, and action plans; and enhancement of communication through the organization of annual meetings of secretariats to consider coordination of activities, rationalization of resources, examination of problems of mutual concern, and establishment of priorities for action in areas where the basic objectives and activities carried out under existing instruments are similar or closely linked.

Tolba reminded the group that the growing number of processes to which biotechnology can be applied to satisfy human needs and aspirations for sustainable development makes it an area of global interest. The socioeconomic and environmental impacts of biotechnology require thorough investigation, and any applications contemplated must take into account risks as well as benefits. In order to balance socioeconomic and environmental risks associated with biotechnology application and ensure its prudent management, mechanisms are required to facilitate the development, transfer, and application of modern biotechnology to solve the problems of particular concern to developing countries; to establish effective cooperation with reciprocal benefits between biotechnology-rich developed countries and gene-rich developing countries; and to anticipate the possible negative impacts of biotechnology and develop appropriate national and international regulatory measures. Access to genetic resources and biotechnology inevitably involves the problem of intellectual property rights. The legal instrument would address only the technologies that would improve the conservation, rational use, and sustainable development of biological diversity.

Tolba listed the following major issues as needing clear definition: The objectives of a convention on biological diversity; the legal status of biological diversity as a basis for subsequent rights and obligations of states; state sovereignty over the elements of biological diversity; equitable sharing of costs and benefits deriving from conservation and use of biological diversity; access to in situ and ex situ biological resources and to relevant technologies; the compatibility of biological diversity conservation and sustainable development; the role of the range of existing legal instruments on various aspects of biological diversity; special needs and interests of developing countries; and financial and institutional mechanisms to provide for equitable cooperation of various groups of countries.

During the meeting the working group had to take into consideration not only the note outlined above, but also a number of studies presented by subcommittees and a presentation of suggested elements for a global framework convention that had been prepared by Rummel-Bulska in collaboration with representatives of some other members of the Ecosystems Conservation Group (FAO, UNESCO, IUCN, and WWF).

Regarding financial arrangements, the group agreed that uncertainties about the total cost of the project should not delay development of the new legal instrument. More accurate country information would help refine these estimates, taking into consideration the full range of biological diversity and ecosystems, not only those found in tropical ecosystems, and the costs of technology transfer as well as of finding alternatives to activities that threaten biological diversity and sustainable development. Although governments were already investing considerable sums in national conservation activities and multilateral and bilateral donor agencies were contributing to conservation of biological diversity, these flows were inadequate to meet in a timely and satisfactory manner the basic conservation needs identified by the working group; neither should the additional funds required for developing countries to achieve conservation of biological diversity be diverted from current development programs. The Global Environment Facility proposed by the World Bank, UNDP, and UNEP could be considered as an element of a funding mechanism. Contributions might be provided by the parties as an assessment based on the industrial and commercial exploitation of, or trade in, genetic resources. The delegates agreed that a pilot financial mechanism was needed pending a better understanding of the costs and benefits of the program. It should combine funding and clearinghouse mechanisms including support to priority conservation needs identified by the working group at its second session and should be under the supervision of the contracting parties, who would closely coordinate with existing funding institutions. The delegates affirmed that free access to biological diversity or biotechnology does not mean free of charge.

The working group also took up the issue of biotechnology transfer, an important element of the planned legal instrument with the potential to contribute to improved conservation and sustainable utilization of genetic diversity. However, because of the complexity of the issue, it was difficult for the group to reach firm conclusions at that third session on how and to what extent the biotechnology issue should be addressed in the convention; the idea of establishing an information clearinghouse for required biotechnologies as proposed in Tolba's statement would be reexamined at the next meeting of the technical working group; similarly, there was merit in further consideration of many of his recommendations.

The working group urged that its concerns on the close relationship between access to genetic resources and biotechnology be brought to appropriate fora such as GATT or the World Intellectual Property Organization (WIPO) to sensitize those groups to issues related to biological diversity. It is important that states bear in mind issues related to conservation and sustainable use of genetic resources when dealing with intellectual property rights issues in fora where these are currently being dealt with. The potential value of natural resources should be recognized, and owners should receive appropriate compensation. A balance should be struck between the legitimate rights of germ plasm owners and technology owners, recognizing the needs of both.

The general agreement reached during the three meetings of the technical working group was that

1. an international legal instrument without firm commitments to funding would be meaningless;
2. those who enjoy the greatest economic benefits from biological diversity should contribute equitably to its conservation and sustainable management;
3. a new partnership should be developed and funding for developing countries should be characterized as cooperation among countries;
4. in providing sufficient new and additional funds in a spirit of common responsibility, the costs of conservation should not fall disproportionately on countries with significant biological diversity; and
5. there was a need to incorporate an innovative mechanism that facilitates access to resources and new technologies, including those in the private sector, and for this reason, whatever the final financial arrangements,

there was a need for a special fund management mechanism as part of the international legal instrument.

By July 1990 the technical working group had confirmed the view of the UNEP and its close partners, the FAO, UNESCO, IUCN, and WWF, that a global convention was needed and had identified all the elements to be included. Meanwhile, the FAO and IUCN had already drafted a number of articles for inclusion in the convention. Rummel-Bulska—with the assistance of a small number of experts essentially from the IUCN and FAO—put all the elements into legal language, making every effort not to change the language presented by governments. A document was prepared containing elements for possible inclusion in a global framework convention on biological diversity. It was circulated to governments and formed the basis for discussions at the fourth meeting.

The working group met again in November 1990, a meeting attended by seventy countries, fifty of them developing. This time the representation was by legal and technical experts, many of them hoping to begin negotiations; with the bureau of the previous sessions staying on temporarily, this seemed for a time possible, but it was not to be. Several experts were very reluctant to begin development of a convention for fear of entering into lengthy and complicated negotiations, developing countries were not interested in a convention that addressed only the conservation of wildlife, and developed countries were concerned that they would be asked for additional financial resources. The participants hotly debated the elements for inclusion in the convention, whose scope had broadened from what was originally expected and now covered in situ and ex situ conservation of both wild and domesticated species; use of biological resources; sharing of benefits accruing from their use; and the transfer of financial resources and of technology, including biotechnology. Forty-three of the seventy countries at the meeting presented the secretariat with written comments on, additions or deletions of, or changes in the elements for inclusion in the draft convention.

The working group failed to agree on either the composition of its bureau or the structure of the negotiations, a disagreement that would seriously hamper the start of its next meeting, in February 1991, at which it was to begin actual negotiations. The group then got bogged down in a competition between Latin American and Caribbean countries, on the one hand, and African countries, on the other, over the chairmanship. The Africans nominated Kenya and the Latin Americans and Caribbeans nominated Chile. All compromise formulas were rejected; members of the working group had to resort to voting, electing Vicente Sanchez, Ambassador of Chile to Kenya, as chairman, and three regional vice chairmen, from Kenya (Joseph Molero), Russia (G. Zavarzin), and Denmark (Veit Koester), and a rapporteur general from Pakistan (Tauher Husain). Once this hurdle was cleared, the meeting organized into two subgroups, one chaired by Molero, Vice Minister of Foreign Affairs of Kenya, and the other by Koester, Director General of the Danish Ministry of Environment. The first sub-group dealt with principles and general obligations, conservation and relation of the convention to other treaties, the second with the sensitive issues of access to genetic resources and to technology, financial resources and mechanisms. For fifteen months positions were adjusted and views were changed, frustrations continued and tension prevailed.

The Intergovernmental Negotiating Committee (INC), May 1991–May 1992

In May 1991 the working group met for the second time under the new bureau. The meeting of the previous November had made evident the deep disagreements and suspicions that would accompany the upcoming negotiating sessions, which were at several points very difficult, and there arose widespread feelings that the negotiators would never reach agreement on a convention in time for its signature at the Earth Summit, scheduled for June 1992. This anxiety would persist until 5 P.M. on the last day of the plenipotentiary conference, but the final hours of the day saw a breakthrough, and before midnight of May 22, 1992, the convention was adopted.

In opening the May 1991 session of the working group, Executive Director Tolba again took the active role that he made the hallmark of the sponsoring organization. He stated that although convention building is a painstaking and exacting process, he was sure the delegates would succeed, because they saw eye to eye on several pivotal issues: All countries,

North and South, accept the fact that global environmental problems have ushered in a new sense of global partnership. All participants agree that a convention on biodiversity must respect the inalienable sovereignty of states over their natural resources; no one wants to create a paper tiger, but rather to build a strong, clear, and concrete agreement, with costed commitments to action. Although it was hoped that the task would be completed in time for the 1992 conference in Brazil, no one was prepared to sacrifice content for expediency. All agreed that the cost of conservation could not fall disproportionately on countries rich in biological diversity, and that there must be fair compensation for access to biological resources and to relevant technologies. Finally, the loss of biological diversity is not only an environmental issue; it is also a development concern, affecting industry, agriculture, forestry, medicine, and other priority areas. Tolba then went on to stress four issues that he considered of particular importance, taking a strong position to which he still adheres.[1]

In the general debate during that session, a number of delegations pointed out that for developing countries their enormous external debt and ever expanding population growth would become a crushing burden if they were to carry out conservation activities. If they were to assume the responsibility for implementing the convention on biological diversity, the debt problem should be examined and adequate additional financial resources provided. The committee's deliberations emphasized the need for

1. further rationalization and coordination of existing international legal instruments on biological diversity;
2. equitable distribution of resources between developing and developed countries and sharing the global responsibility of conservation;
3. establishing a relationship between conservation, utilization, and property rights;
4. building the new international legal instrument on existing international legal instruments and measures, taking into consideration the efforts of other international and regional organizations within and outside the United Nations;
5. raising public awareness by appropriate education programs at all levels;

6. applying measures to curb the contamination of the biosphere by pollutants;
7. establishing inventories of flora and fauna at the national level; and
8. development and implementation of action programs and other measures for conservation of biological diversity pending preparation of the legal instrument.

The group also agreed that whenever the concept of common heritage was referred to, it did not mean the establishment of collective international rights to resources within national jurisdictions, nor did it infringe on the permanent sovereignty of states over natural resources.

The third session of the negotiations was held in Madrid in June–July 1991, with the government of Spain making every effort to provide an environment conducive to positive dialogue. By this time the working group had been renamed the Intergovernmental Negotiating Committee (INC). The Madrid meeting was the real start of negotiations and saw the unfolding of deep differences between negotiators from North and South and among negotiators of each of the geographic groups. In the three months between the second and third sessions, several notes were prepared by Tolba and Rummel-Bulska, with the support of their colleagues on the secretariat, and circulated to governments and observers. In spite of the press of responsibilities for organizing the meetings, they also drew up and circulated a revised version of the draft convention. This was made possible by the commitment to the environmental cause of everyone from the UNEP secretariat involved in the process of negotiations: Ruben Olembo, Hamdallah Zedan, Mona Bjorklund, Alexander Timoshinko, Suzan Bragdon, and a few junior officers.

The revised draft convention included all the draft articles and was reduced from sixty pages for half the convention to thirty-three pages for the whole convention except for definitions. Revising the draft convention fell to Rummel-Bulska, who achieved it by eliminating repetitions, comments, and cross-referencing. This was the first real consolidated draft convention, previous texts having been more akin to groups of elements for possible inclusion in a draft convention. It had a number of alternatives under some articles and some seventy bracketed items, mostly relating to access to finances and modalities for technology transfer. It was reviewed in April 1991 by a small group of international lawyers from various regions of the world.

UNEP Executive Director Tolba reopened the third negotiating session by reminding the delegates that although there were no fundamental disagreements concerning the broad goals of the convention, with the target date only eleven months away, progress at the present session was crucial. He called for a drastic narrowing of the numerous options in the revised draft convention, reiterating, however, that content could not be sacrificed to expediency. Obligations should include in situ and ex situ conservation; intergenerational equity and responsibility; arrangements for the transfer of technologies, including biotechnology; and the establishment of financial mechanisms, with a proper balance between national sovereignty and collective responsibility. A key point to resolve concerned the wordings "free access," "fair access," and "equitable access." He drew attention to the importance of the link between the convention and the discussions over intellectual property rights in the then-current Uruguay Round of negotiations within the GATT.

Emphasizing that a means of evaluating biological resources had to be found that would recognize that their loss was irreversible, he suggested that a small, regionally balanced group of economists and biologists should be set up to prepare a note on this issue for the next session of the negotiating committee.

There were in the draft convention two options for financial mechanisms, a multilateral trust fund with an initial base of US $500 million and an international corporation with initial funding of US $200 million; two model financial mechanisms, the Global Environment Facility and the Multilateral Fund for the Implementation of the Montreal Protocol, were described. It was important to reach agreement on an order of magnitude for the finances needed during the first few years of operation of the convention.

The final issue to be resolved was the degree of detail that the convention should contain. On some issues it was essential to be specific; on others it was preferable to agree on general principles and leave the details to protocols. Tolba felt it would be inadvisable at that stage to debate what should go into the convention and what into protocols—an issue that was expected to resolve itself during negotiations.

Sanchez, the chairman of the committee, pointed out in his opening address that the basic problems concerned methods of in situ and ex situ conservation, access to genetic resources and technology, and the necessary financing. For the most part, genetic resources were concentrated in the developing countries and access to them had been relatively unrestricted, whereas access to the technologies needed to exploit them had been protected by intellectual property rights. Accordingly, any measures taken to protect biological diversity had to be accompanied by efforts to make both the necessary resources and the technologies readily available. Internationally valid agreement must be reached on methods of using such resources, including mechanisms that would provide the countries and communities of origin with an equitable share of the benefits. He further indicated that the recent inclusion of intellectual property rights in the GATT negotiations might give rise to complications that the committee should look into. Another problem that deserved attention was the failure of market mechanisms to cope with the conservation and use of genetic resources. Market forces alone did not provide sufficient incentive for socially desirable levels of investment. Furthermore, although there could be no doubt that biotechnology could be a useful tool for the conservation of biodiversity, its use to introduce improved species had frequently led to a loss of biodiversity. Lastly, the provision of funds for the application of the convention, nationally and internationally, was of fundamental importance. It was to be hoped that those who were well placed on the world economic scale would make generous contributions, as some had already done in the past.

These two statements set the tone for the negotiations. Yet right from the beginning polarization was obvious. There were also several delays for reasons having nothing to do with the substance of the negotiations—late translation into one or another of the six official UN languages used; a misunderstanding of a delegate's statement because of inaccurate interpretation; even whether this was the third or the first negotiating session of the Intergovernmental Negotiating Committee. Sensitivities and tensions increased, especially between the developed and developing nations.

The issues of country of origin of biological resources, technology including biotechnology, and financial resources became increasingly contentious, as may be seen in three paragraphs of the report on the session:

One group wanted the phrases "country of origin" and "country providing genetic material and/or genetic resources" to be defined; several delegations wanted to clarify the phrase "utilization/use of biological diversity" and to note in an appropriate place in the convention that "technology" included "biotechnology," avoiding the need to place "including biotechnology" after every reference to "technology." The issue of what was meant by the phrases "adequate," "new and additional," and "new and additional financial resources" kept arising, and the secretariat was requested to prepare a note on the interpretation of these phrases.

A sense of the tension that prevailed may be conveyed by the following exchange. With regard to the issue of technology transfer, the representative of the Netherlands, speaking on behalf of the European Community and its member states, said that when discussing that issue it was important to start from a substantive basis, including documents that were presented to other fora, and that the discussion should include elements such as training, education, institutional aspects, diffusion of technology, or even commercialization of technology. The description of relevant technologies also needed to be improved. His statement was supported by some delegations; others understood it to mean that discussion of Articles 15 and 16 of the draft convention dealing with transfer of technology should be postponed until after the documents referred to had been discussed. The Group of 77 and China said that Article 14 relating to access to biological resources was completely open for further changes of a substantive nature, regardless of the outcome of discussions on Article 15 relating to access to technology or any other article, and furthermore that they would consider the outcome of discussions on Articles 15 and 16 and other articles in order to harmonize them with the spirit and content of related articles. In response to this the representative of the Netherlands stated that he had not intended to imply that the working group should not or could not discuss technology transfer; the EC member states were certainly willing to discuss the text paragraph by paragraph if the group so wished.

The revised draft convention presented to the negotiating committee at the start of the session contained 39 articles; ten days later the committee had partially considered only nine of them. Of course, these articles covered the most contentious issues: objectives, fundamental principles, general obligations, access to biological resources and to technology, and financial needs and mechanisms. As they emerged from the committee the nine articles included two additional ones relating to exchange of information and handling of biotechnology and distribution of its benefits. The entire article describing the objectives of the convention was bracketed, and the rest of the draft convention had more than 160 brackets, compared to 70 for the 39 original articles. In spite of its discouraging aspect, this was a sign that governments had begun serious negotiations; these brackets indicated the magnitude of their differences. Now the pace had to quicken. There were hardly eleven months left before the date assigned for signing the convention. The committee agreed to the proposal that they should meet in September and November 1991 and in February and May 1992. They would then revise, review, translate, and distribute documents for the following meeting, prepare notes, and above all start the process of informal consultations between those with widely disparate positions, of which there were many, to try to narrow the differences. These additional meetings placed a very heavy burden on the Secretariat, particularly on Director Tolba and Dr. Rummel-Bulska.

During the month that followed the Madrid session, Tolba and Rummel-Bulska reconvened the small group of lawyers to review the articles that had not been considered in Madrid. A second revised draft convention emerged and was circulated to governments. Papers were prepared by the authors for the fourth INC session, one clarifying interpretations of various terms and another discussing financial resources and property rights.

The fourth session of the Intergovernmental Negotiating Committee took place in September–October 1991. It was attended by eighty-one countries, fifty-seven of them developing, and a large number of UN organs and NGOs attended or intensified their presence.

In opening the session the chairman stressed the increasing gulf between rich and poor and the need for a new style of negotiating because the conservation and rational use of biodiversity is the collective responsibility of all. Reminding the delegates that important changes regarding intellectual property systems were being proposed in GATT, he urged national delegations to ensure that their approach to the various negotiations was consistent, because it was difficult to reach agreements, and unnecessary tensions were produced when there were differing standpoints from the same government in different fora.

The capacity of the Consultative Group on International Agricultural Research (CGIAR) to patent research results meant that it would collaborate only with countries having appropriate intellectual property protection systems and furthermore that it appeared that consideration was being given to the notion that research centers should be free to sell their genetic material in the private sector without sharing profits with the providers of the material. Some private-sector industries had proposed that if biotechnologies were transferred to developing countries, the latter would only be allowed to market their production locally; that would be a disincentive to developing countries to acquire biotechnology. The situation was further aggravated by the sale of substitutes for natural products by private companies.

Until recent times, local communities had been the users and custodians of biological richness. Their knowledge and rights should be respected to ensure that the convention was firmly rooted. Indeed, many people considered it unfair that biodiversity should be seen as the common heritage of mankind. Inequalities and imbalance had to be remedied in order to achieve a more stable new world order and to ensure that democracy prevailed in connection with natural world and trade relations, as well as in the political order.

Tolba's opening statement reminded the negotiators that they were expected to build a meaningful, flexible, and fair convention to bring to Brazil, and their tight timetable could not create any shortcuts or compromises that sacrificed content. He stated that progress could not be made regarding additional financial resources and technology transfer until there was a consensus on two basic questions. First, the question of value: The viability of the present economic system was increasingly dependent on access to biological resources, yet the means of assessing the value of biodiversity was lacking. Economic systems were as yet unable to recognize the value of the unknown or undiscovered. The second question was that of technology: Progress was measured in terms of development and use of sophisticated technologies, yet the way in which new technologies were regulated hindered their dissemination where they were most urgently needed.

In spite of all this prodding, almost next to nothing was achieved during this negotiating session. The first subgroup barely reviewed two articles and one paragraph of a third and left them with several brackets. Working Group II simply reached an understanding that was to form the basis for further negotiations. The issue of technology transfer turned out to be a stumbling block, with widely divergent positions by developed and developing countries. The subgroup requested the Secretariat to present a note on the interpretation of the terms "fair and favorable"; "fair and most favorable"; "equitable, preferential, and noncommercial"; and "preferential, noncommercial, and concessional." Out of by now more than forty articles constituting the draft convention, thirteen were discussed during the ten-day meeting. They dealt with implementation measures; in situ and ex situ conservation; traditional, indigenous, and local knowledge; access to technology; exchange of information; transfer of technology; technical and scientific cooperation; and international cooperation. Completely new drafts were presented at the session. What emerged on October 2, 1991, barely eight months from Brazil, was that five articles out of the thirteen discussed and eight paragraphs of the remaining articles were bracketed, constituting almost half the text from the end of the fourth negotiating session. In the remaining part, there were more than 150 brackets around words, phrases, and sentences.

The fifth session of negotiations began in December 1991, seven weeks later. The revised draft presented to the session contained forty-six articles, fewer than fifteen having received first or second readings. None had been adopted even at the subgroup levels. The articles on objectives, definitions, identification, and monitoring; on the situation of developing countries; on traditional, endogenous, and local knowledge as well as almost all of those on general obligations, were fully bracketed. Now the full text had more than 300 brackets. Assuredly, some of them were around commas and periods and prepositions, but some were around very crucial issues in the negotiations. For example, phrases were bracketed that dealt with the fair and equitable sharing of the benefits of research in biotechnology arising out of conservation of biological diversity; providing adequate, new and additional funding to the developing countries; the transfer of technology to developing countries on preferential and noncommercial terms; and assurance that activities within states' jurisdiction or control do not cause damage to the biological diversity of other states or of areas beyond the limits of national jurisdiction.

At the beginning of the December 1991 session, in which seventy-five countries participated, the chairman and the executive director made very brief statements stressing the short time left and the large number of issues yet to be resolved. The delegates agreed that no statements were to be made; rather they would plunge straight into negotiations. Yet at the end of the session, Group I had considered ten articles, fully bracketed four of them, and failed to agree on any; Group II had considered four articles and disagreed on all of them. These fourteen articles emerged with more than 120 brackets. The basic issues were still far from being resolved. The positions of the North and the South, as reflected in the alternative bracketed languages, were very far apart.

That was the situation at the end of 1991, less than six months from the time set for signing the convention. It was abundantly clear that informal consultations must begin. Chairman Sanchez, Tolba and Rummel-Bulska started a series of informal discussions with individual countries and small groups of countries in preparation for the sixth round of negotiations, in February 1992.

A collateral problem had developed: whether or not the document would be called the Nairobi Convention. The Brazilians were adamant that it should not, since it was to be open for signature in Rio de Janeiro during the UN Conference on Environment and Development, and in light of this they raised the question of whether there should be a separate plenipotentiary conference to adopt the convention or whether this would be better left to the last session of the INC.

The Kenyans very much wanted it to be called the Nairobi Convention, because most of the negotiations had been carried out there. Each party made its position clear during the plenary sessions, that of Brazil supported by a number of Latin American countries. The issue stayed alive through the whole negotiating session. Ultimately a formula was proposed that, though it made no one happy, was accepted by both sides, to have a plenipotentiary session in Nairobi to adopt the agreed text of the convention and the resolutions as recommended by the INC, at which time declarations by governments would be made. The report of the conference together with the adopted text of the convention and the resolutions of the conference, and the declarations made by governments at the time of adoption of the agreed text, would constitute the final act of a treaty to be called the Nairobi Final Act. The convention itself would carry neither the name of Nairobi nor of Rio de Janeiro, and that is what happened. So one more problem—in fact, a big problem, although it had nothing to do with the substance of the convention—was solved.

In February 1992, the INC reconvened and started negotiations with a revised text of the convention containing some 350 brackets. The sixth session was attended by eighty-three governments, the highest number ever in attendance. Once again a large majority, fifty-eight countries, were from the developing world. By the end of the session it was becoming increasingly obvious that governments were still very far from reaching agreement. The situation became very delicate; with three months remaining until Rio, the convention was in bad condition. The article on its objectives was completely bracketed, and there were six brackets covering fair and equitable sharing of benefits, adequate new and additional funding, cost and benefit sharing between developed and developing countries, and favorable economic and legal conditions for technology transfer on preferential and noncommercial terms. These same issues had also been causes of contention between developed and developing countries during the negotiations of the Montreal Protocol on Substances That Deplete the Ozone Layer, the Basel Convention on the Transboundary Movement of Hazardous Wastes and Their Disposal, and the Climate Change Convention. They had figured prominently in the UNCED preparations as well as during the conference itself and surfaced clearly after the conference during the negotiations for the desertification convention and the meetings of the UN Commission on Sustainable Development.

Again at the end of the session the article on use of terms was fully bracketed, as well as that on fundamental principles. The article on general obligations was also fully bracketed, with a complete alternative text also between brackets. Five more full articles were bracketed and several had disputed alternatives to several paragraphs. The negotiating session ended with a draft convention having more than 250 brackets—down from the some 350 brackets at the start. The issues that remained in dispute included the rights of countries or indigenous people providing genetic resources; conditions for the transfer of technology and for access to genetic resources; national jurisdiction or control; additional protocols; what is to be conserved in situ and ex situ; the link between implementation of the provisions of the convention by developing countries and providing them with technical and financial resources; impact assessment of incentive measures; global lists of threatened species and ecosystems; handling of biotechnology and distribution of its benefits; financial resources and mechanisms; the secretariat; and several other issues. With but a single meeting left before the date set for signing the convention, the situation appeared bleak.

The seventh and final negotiating session was scheduled for May 1992, barely a week before the Rio conference. Tolba, Rummel-Bulska, and the rest of the UNEP members of the secretariat—Ruben Olembo, Alexander Timoshenko, Hamdallah Zedan, and two junior staff members in particular—set themselves the seemingly impossible task of drafting compromise formulations for all that was in brackets, consulting with the chairman and key delegations through all means of communication. A paper was prepared. Chairman Sanchez introduced a number of proposals, and finally a joint formulation was proposed by Sanchez and Tolba, covering twenty-seven articles and a preamble. The proposed convention was presented to the negotiators on May 11, 1992, the first day of the last negotiating session, less than four weeks from the Rio Conference, the target date for its signature; many of the delegates were pessimistic.

While the delegations worked on the basis of the joint paper, Sanchez and Tolba held, separately and jointly, nonstop informal consultations with individual delegations or groups of delegations that continued until the early morning hours. Koester, who chaired Group II throughout, did an outstanding job in trying to smooth out the differences over the most difficult issues: access to biological resources; access to technology including biotechnology; financial resources and mechanisms; and global lists of threatened ecosystems and species. These issues continued to be the subject of negotiations and intensive informal consultations among the delegations themselves in addition to those carried on by Sanchez and Tolba.

The May 1992 session of the Committee, attended by a record 101 governments and 25 United Nations and other international organizations, came on the heels of the last session of the Negotiating Committee on the Climate Change Convention, adopted in New York. The developing countries felt, in general, that they had been put under pressure to agree to the final text of the Climate Convention and remained very tense at the last negotiating session for the Biodiversity Convention.

In his opening address, Chairman Sanchez drew the Committee's attention to the fact that although there were still several bracketed sections in the text, many paragraphs and articles were already clear of brackets, which meant that agreement was very much achievable. He welcomed the adoption of the Convention on Climate Change and expressed the hope that the Biological Diversity Convention negotiations, which had started well before those on climate change, would also be able to reach a successful conclusion. He referred to the informal note prepared by the UNEP executive director and himself to assist the countries in the negotiations and advised them to make use of it.

Tolba then pointed out issues needing compromise solutions. Some parts of the article on the use of terms had to be further developed on the basis of scientific evidence; it might be advisable simply to establish the principle of global lists of areas, processes, and activities, leaving it to the parties to decide on the basis of scientific advice when to establish them and what their contents should be; although the principle that there was need for new and additional resources for developing countries and countries in transition no longer seemed to constitute a problem, divisions still existed among negotiators on a number of issues relating to financial resources and mechanisms. Two issues on technology transfer needed to be resolved. The first concerned the very complicated and controversial area of intellectual property rights; if this issue could not be settled at the current session, the Committee could perhaps move forward where consensus was possible and find a mechanism for resolving the issue in the future. The second was that of biotechnology, its transfer, information regarding the introduction of its products and the sharing of profits from its applications to biological resources. Tolba emphasized that instead of trying to develop a perfect solution for everybody, the session should set in motion a process out of which would grow solutions advantageous to all.

Negotiations continued until May 22, the last day of the meeting. On the morning of that day there were still a large number of brackets and disputed issues. The three leaders of the INC held a series of informal consultations with the differing parties, which resulted in compromise formulations. Sanchez, on behalf of Tolba and himself, introduced the resulting amendments at 3:00 in the afternoon, stressing that they represented a compromise and that the draft convention, as amended, would be considered by the INC as a package that should be preserved. There was a brief silence, and then, at 4:11 P.M., the delegates burst into applause. The miracle had been achieved. The amended text was accepted, although even at that moment various governments made statements[2] showing how much they differed in their assessment of the results achieved.

Two hours after the conclusion of the work of the INC, the Plenipotentiary Conference was opened to adopt the agreed text of the convention and four resolutions: one designating the Global Environment Facility as the interim financial mechanism for the convention, two dealing with urgent activities that must be taken in line with the provisions of the convention and pending its entry into force and the fourth paying tribute to the government of Kenya, the host of the Plenipotentiary Conference. The conference also listened to declarations by governments. All this was included in the final Act, which was signed in the final hour of the last day of the meeting.

At the time of adoption several delegations again made declarations. Those of France, India, and the United States were quite significant. France was concerned over the failure of its proposal to establish global lists of biological resources, India over the issues of liability and compensation and their relation to other international agreements and the financial mechanism, and the United States and most of the developed countries had exactly the opposite concern and were deeply concerned over the issue of intellectual property rights. Therefore, a consensus was reached over a text of the convention that pleased no one. This seems a good indication that the provisions of the convention were balanced.

The convention was opened for signature at a Plenipotentiary Conference on the Convention of Biodiversity convened in Rio de Janeiro at the beginning of UNCED, the UN Conference on Environment and Development, in June 1992. Since the convention had been adopted in Nairobi, the U.S. administration was completely negative toward it. President Bush personally took issue with it, during his election campaign declaring on television that it would affect, presumably negatively, every family in the United States. There seems to have been a misunderstanding or misinterpretation of some of the provisions on biotechnology, genetic material, and access to technology. Although the United States refused to sign the convention at Rio, 157 governments did sign, mostly at the level of their heads of state. The Convention must have been quite balanced to have met with the approval of the great majority of the countries attending the Earth Summit.

Lessons Learned

The negotiations on the Biodiversity Convention spanned a period of more than five years. During this time, while no one disputed the fact that the loss of genetic resources, ecosystems, and species was accelerating at an alarming rate through human actions, when it came to adopting measures for halting this acceleration and trying to reverse the trend, negotiations centered on political, financial, and economic gains, and most governments had their own agendas, widely divergent and difficult to bring together.

On the other hand, the discussions of technical and scientific facts were calm and relaxed, and agreements were most easily achieved at these times. When the science and economics of the issue were clear and unwavering, when the overall objectives of the convention were clearly spelled out, and when there were enough strong personalities to lead, to speak with authority, and to seek compromise, a good deal of agreement could be reached. The negotiations greatly benefited from the presence of a number of delegates, among whom were Julio Barboza (Argentina); Hugh Wyndham and Kathy Leigh (Australia); Donald Cooper (Bahamas); Marco Azambuja, Luiz Felipe-Soares, and Julio Bitelli (Brazil); Arthur Campeau (Canada); Li Song (China); Vicente Sanchez (Chile); Garcia Duran German (Colombia); Veit Koester (Denmark); Michael de Bonnecorse (France); Wolfgang Hoffman (Germany); Avani Vaish (India); Dalindra Aman (Indonesia); Lian Wen Ting (Malaysia); Juan Antonio Mateos (Mexico); Leon Mazairac (Netherlands); Zavarzin (Russia); Ulf Svenson (Sweden); Fiona McConnel and Patrick Szell (United Kingdom); and Eleanor Savage (United States). It was also important that the representatives of the organization serving the negotiation of the treaty took an active but objective stand, defending the rights of the environment without trespassing the limit of tolerance of the negotiating governments.

During these negotiations and those on climate change, new terms entered the vocabulary of environmental negotiations:

- the common concern of mankind as a different concept from the common heritage of mankind
- common but differentiated responsibilities
- burden sharing among developed countries
- intergenerational equity and intergenerational responsibility
- the rights of indigenous communities—in this case, in sharing the benefits of using the biological resources that they have bred over the years.

Notes

1. As presented at the time, the statement still appears to the senior author to be objective and well balanced:

"First: there was a basic consensus that common concern over the conservation of biodiversity requires the participation of all countries and all peoples in a global partnership. It implies intergenerational equity and fair burden sharing. The common concern calls us to strike a balance between the sovereign rights of nations to exploit their natural resources, and the interests of the international community in global environmental protection.

"Second: financial mechanisms: the draft convention must contain concrete and binding commitments to funding. A key challenge is to consider how to provide financial support for the protection and sustainable use of biological diversity. Initial cost estimates suggested one to ten billion dollars are needed yearly over the next 10 to 15 years to meet conservation costs for the priority areas identified by the Ad Hoc Working Group of Technical Experts during its second session in February 1990.

"Imperfect knowledge about financial requirements cannot be an excuse for procrastination in defining financial modalities. The convention must agree on additional and sustained financial resources. We need to examine how these new and additional financial resources will be generated and equitably distributed. Should they come from the thin budget lines of overdrawn treasuries? Should we assess alternative revenue generating sources, such as user's fees, biodiversity charges or fiscal policies that promote biodiversity conservation and sustainable utilization?

"Third: the interlocking issues of availability and access to biological resources and to relevant technologies: Access to biological resources and the availability of biotechnology and other technologies relevant to the rational use of biological resources are complementary and inseparable. Sovereign states expect and should receive fair compensation for the use of their genetic resources. And the private sector—which invests millions in research and development for new technologies—expects and should receive fair compensation for participating in technology transfer arrangements, in supporting education and training and in developing indigenous technologies.

"There is a continuous argument in international fora that because technology patents are held by private firms, governments cannot, in view of current international treaties, dictate to the private sector policies regarding transfer of patented technologies. By the same token, areas rich in biodiversity are mostly in private hands. If governments of developing countries are expected to convince landowners to participate in the implementation of provisions of the convention that require access to biological resources, then it is certainly not asking the impossible for industrial countries to persuade their private sector to act in a similar way."

2. Australia stated, "I believe we have done something worthwhile and valuable in the convention and we will realize this when the fatigue and tension of the last

few days have passed. This document presents a very delicate political compromise. Because of the need to find those compromises, some of the resulting language is less than ideal from some points of view. As to the complex financial provisions, these will have to be carefully examined and implemented in a way which respects the rights of members of other organizations with which the conference of the parties established by this convention will deal and of the organizations themselves."

Austria had reservations in particular with respect to the provision defining the relationship between the conference of the parties and the institutional structure. Austria shared in this respect the views expressed by the speaker of the European Community.

The Bahamas pointed out that the Biodiversity Convention is not perfect from the point of view of any delegation. However, as an instrument for the conservation and sustainable use of biological diversity, it is a major accomplishment.

Canada said the Biodiversity Convention is an important achievement in international environmental law. It contains a broad range of measures that it is anticipated will serve as a framework for conserving our natural heritage.

Chile said: "We reached the end of this period with very positive agreements, the extent of whose consequences we cannot yet foresee."

The Chinese delegation commented that the developed countries would be the first to take actions in order to promote the realization of the goals of this convention, and that the fulfillment of the obligations stated in the convention by the developing countries will depend on whether the developed country parties comply with the stipulations of transferring technology and of providing adequate new and additional financial resources.

The Colombian delegation said that although the final result of the negotiating process was not fully satisfactory for them, in a spirit of agreement and understanding they joined in the consensus, in the sure knowledge that it opens the road toward conservation and sustainable use of biological diversity through the future strengthening of the treaty.

Ethiopia expressed its dissatisfaction with the provisions protecting patents and other intellectual property rights without commensurate regard for informal innovations, which opens the way for use by countries with the technological know-how of genetic resources and innovations from countries without the know-how in patents and other intellectual property rights and for taking them out of reach of even those countries that created the genetic resources and innovations, and yet agreed to go along with the convention with the express wish that, at a later date, these concerns will be addressed.

France considered that one article can in no way be interpreted as affecting the general provisions governing international state responsibility and that the provisions of another article can in no way mean that the amount of funding from each contracting party is determined by decisions of the conference of the parties. The French delegation protested against the manner in which the text was adopted and considered that the chair seemed to have gone beyond its rights. In this respect it must be noted that France was the country that proposed an article on global lists and one on the procedures to prepare them. All efforts to keep any reference to global lists were vehemently objected to by a number of developing countries, led by India, and the two articles were finally deleted. This explains some of the harsh language of the French statement.

Ghana said it had reservations respecting the balance in property rights as it appears in this convention. But in spite of these reservations, the subject that this convention deals with merits its serious consideration and support.

Japan made a lengthy statement that covered the financial mechanism under this convention, the modalities of contribution and the necessary procedures to clarify these points.

The delegation from Kenya said that they feel proud and satisfied that this has happened in their country and in Nairobi.

Malaysia stated that although concurring with the consensus on the article of the convention dealing with handling of biotechnogy and distribution of its benefits, it understood the term "living modified organisms" to mean genetically modified organisms. It expressed reservations regarding the terms of transfer of technology in the convention as well as on interim financial arrangements.

The Swiss delegation stressed the progress accomplished in establishing the framework conditions for cooperation among states.

The United Kingdom had serious reservations about the financial articles.

The delegation of the United States was quite critical. They said that the United States did not intend to disrupt its existing federal and state authorities. Rather it is committed to expanding and strengthening these relationships. Should the United States become a party to this convention, its intent would be to meet its conservation obligations through existing federal laws, and it would look forward to continued cooperation with the various states in this regard. They reaffirmed their belief that the impressive gains in world food production have been a consequential result of the free flow of genetic resources among all countries and indicated that the United States intends to retain its policy of open access to genetic resources and related technologies. They cautioned that if the application of this convention restricts access to these resources, the world will not benefit from this instrument. They also stated that the United States strongly supports the promotion of the transfer of technology and scientific cooperation, provided that intellectual property rights in technology are recognized and protected.

9

Making International Environmental Agreements Work

Unfinished Business

Over the past two decades a number of international environmental agreements, treaties, and conventions have been achieved. Beginning with the 1972 United Nations Conference on the Human Environment in Stockholm, environmental awareness and responsibility have produced significant measures to control environmental degradation due to human activity. In 1992 a conference in Rio de Janeiro produced Agenda 21, which summarized the environmental and developmental agreements achieved by the international community since the Stockholm Conference. The six instruments whose negotiations have been described in this volume are only a few of these, but they are representative of the international environmental agenda that has evolved since 1972, and the problems that beset their establishment are typical of those faced by all the others. Some lessons learned in earlier negotiations have been useful to latecomers in the field; others seem to have been forgotten; and new issues present new challenges. The six examples cited were chosen because of our close personal association with every step of their negotiation.

The authors have a cumulative total of more than forty years of service to the United Nations environmental community. We have seen the growth of environmental awareness and the results of environmental ignorance in many settings, from Kenya in the 1970s to Europe, North America, and other regions in the 1980s and 1990s. Although we believe there is no cause for complacency, there have been encouraging signs. The community of nations has demonstrated its readiness to agree to binding international and regional legal instruments, from saving the ozone layer, to wetlands preservation, to control of the trade in wildlife. Development assistance agencies generally accept that a healthy environment is a key factor in development. Public awareness, slight in the early 1980s, has grown tremendously, and even more encouraging, our young people support environmental causes. Like the generation before them, who literally grew up with the environmental movement, they are aware of many of these issues, and we may be confident they will provide leadership in the future. Additionally, there is evidence that our environment may not be as fragile as the pessimists suggest, that when care is taken damaged ecosystems can be rehabilitated and in some cases can heal themselves.

But much work remains to be done. There is a growing awareness that transboundary pollution is a dangerous source of international tension. An urgent issue and a source of potential conflict is that of shared freshwater resources. Several ground-breaking agreements have been made to protect the streams, lakes, and aquifers shared by two or more nations, but there is as yet no mechanism to warn of the potential for conflict as populations grow and fresh water becomes scarce or is contaminated. There is a need for research institutions to form teams of hydrologists, water engineers, economists, development planners, political scientists, lawyers, and sociologists to draw up conventions that will protect this vital resource and rationalize its use for the benefit of all concerned.

The end of the Cold War did not bring about the peaceful era expected by many. Erosion of global resources through pollution and waste can only bring about renewed conflicts as people demand food, water, and shelter, and such tension cannot be reduced by applying the fortress mentality of a previous era. The vast amounts spent annually on arms did not stop acid rain from devastating forests throughout Europe, nor the spread of radiation from Chernobyl, nor the chronic pollution problems that plague the Mediterranean and Baltic seas; money spent on nuclear stockpiles would have gone far toward rebuilding the economies of eastern Europe. Many in the defense community have recognized the close connection between environmental health and national security, and they advocate strengthened international cooperation to protect the fragile web of life that is our natural resource base, rather than additional expenditures on arms.

Establishment of global environmental regimes required that new concepts and principles be invoked. These include

- the precautionary principle
- the "polluter-pays" principle
- common concerns of mankind
- intergenerational equity
- new and equitable global partnerships
- common but differentiated responsibility
- public participation
- market-based approaches

In order for these concepts to be useful in future agreements, they must be clarified and firmly defined by a panel of economists, political scientists, and international lawyers.

Economics and the Environment

Current models of economic development fail to consider the value of natural resources. Two centuries ago, Adam Smith, in *The Wealth of Nations*, wrote, "Things which have the greatest value in use—such as water—have frequently little or no value in exchange." Although the environment is embodied in all goods and services exchanged it is not itself exchanged and therefore eludes a market price. The result is that natural resources are treated as "gifts of nature" rather than as productive assets. Economics has great difficulty assigning value to anything outside of mercantile activities. Crops are valued, but the land on which they grow has been all but banished from economic priorities, particularly the neoclassical economic models, and economic development is viewed almost exclusively as a function of capital.

This is an outgrowth of our early experiences in which the environment seemed limitless in extent and in regenerative power. Once a natural resource was depleted, a lake polluted, or a virgin forest exploited, it was easy to move on to new water sources, new soils. Resources were incorporated as a free good in production methods, pollution ignored as a market externality. So whereas we treat industrial plants, machinery, and buildings as productive capital whose value depreciates over time, the natural wealth of nations is not so valued; because we overestimate the natural regenerative capacity of soils, forests, fresh water and fisheries, we dangerously under-value them. It is only when countries begin to assess the task of rehabilitating damaged ecosystems that something like their true value becomes apparent. For example, topsoil eroded from deforested hills has drastically shortened the life of an expensive hydroelectric dam in Costa Rica. A government ministry estimates that the watershed might have been protected at a cost of $5 million; reforestation will cost $50 million.

What is needed is a change in our perception of wealth. Economic practices must be expanded past their present time horizons of monthly mortgage payments, quarterly results, annual income accounts. Natural resource endowments must be included in every nation's inventory of wealth, environmental priorities integrated into every dimension of macro- and microeconomic practices. Ecosystems provide the foundation on which all economic systems are based. Unless economists take a long, hard look at present environmental realities, no economic system will survive.

Putting a price tag on the environment is not a new idea. Keynes' teacher Pigou is credited as the inventor of the polluter-pays approach. In his view, the environment was not a free good. Those who used it, either as a resource or as a sink for waste products, must pay. Affixing a price is difficult under even the most rigorous accounting system. There is as yet no consensus among economists on how to "price" a renewable resource. The phrase "dirt cheap" accurately conveys the general attitude toward soil, but soil is in fact a living community of roots and organic and microscopic life, a storer and dispenser of water, a portfolio of capital that can be conserved in perpetuity. It can also be destroyed in the space of a few years, as U.S. farmers discovered during the Dust Bowl era.

Beyond what might be termed the environmental constituency, there is a rapidly growing body of opinion that we must tread a different path. Germany, France, the Netherlands, Norway, Canada, and Australia are moving toward natural resource balance sheets, as are the World Bank and a few other lending institutions. To set realistic prices on natural capital may require economic instruments like safe minimum standards, selling or leasing of resource rights, and environmental and resource taxes, and incentives such as enlightened subsidies, to encourage industry and commerce to husband resources. In fact, there are three types of capital that must enter into the equation: natural capital, including biogeochemical cycles and species diversity; natural resources, including forests, marine life, topsoil, and air; and man-made capital such as technology. Reforms in economic tools must recognize indicators of national wealth, including natural resource bases, and include means of quickly assessing the impact of environmental decisions on international trade. This is a complex challenge, but it must soon be addressed if we are to pass on to future generations productive opportunities of equal or greater value than our existing capital portfolio.

Trade and the Environment

A number of international institutions are engaged in efforts to establish principles that will guide global trade in the coming decades. Trade policies have a profound effect on environmental protection, and studies have barely begun that will clarify their interaction. When trade policy is invoked to protect the environment, principles of nondiscrimination, minimal application of restrictive measures, and transparency should be applied. Studies of the environmental impact of trade policy as well as of how trade policy is affected by environmental principles such as the polluter-pays and precautionary principles and life cycle management, should be expanded, as should those on the interaction between trade, technological cooperation, and changes in production and consumption patterns. The relationship between the provisions and dispute settlement mechanisms of the multilateral trading system and those of multilateral environmental agreements should be explored. The role of environmental policies as they relate to trade liberalization policies, and the relationship among environmental protection, international competition, job creation, and development deserve further study.

There should be a legal review of the intellectual property rights provisions of the Uruguay Round of GATT as they affect certain environmental agreements, particularly the Biodiversity Convention, and an assessment of the potential consequences of their application on the rights of the people of the countries of origin of genetic resources. The

Agreement on Agriculture of the Uruguay Round will have a profound effect on sustainable agriculture in both North and South. Various national and regional proposals for the use of economic instruments to internalize social and environmental costs, particularly in developing countries, will have both positive and negative effects that need to be assessed. Adequate studies of the above issues should prove or disprove premises embodied in such statements as, "Liberalization of sectors of the economy such as agriculture and health would make many goods and services unavailable to the majority of people," and "Trade liberalization erodes indigenous and diverse systems of agriculture, promoting environmental degradation and reducing nutritive standards."

The Evolution of Policy

Now that environmental treaties have been in place for various lengths of time and enough experience has been gained to evaluate their worth, it would be appropriate to assess their continuing usefulness and their value as models for future treaties. Four issues should be investigated: improving existing environmental treaties and negotiating new ones, consumption and production, transfer of financial resources, and technology transfer.

Existing treaties should receive objective reviews to answer the following questions: Was the treaty drafted with adequate participation of developing countries? If not, what biases need to be corrected, and how? Is there any conflict among the treaty and others in the economic and social domain, such as trade agreements? Here it would be necessary to obtain balanced, objective experts independent of the institutions that developed the treaty. Does the treaty have adequate noncompliance and arbitration systems? Here international lawyers and political scientists would be particularly helpful. Is the treaty adequate to deal with the environmental problem it was designed to address? Are the assumptions used as its basis still valid? This would require input from the scientific community. And finally, the adherence of the parties to the treaty should be assessed and reasons determined for any failure to participate.

A dauntingly long list of issues await action through international agreements. Some of the most urgent are control of the international trade in potentially toxic chemicals; control of land-based sources of marine pollution; application of environmental impact assessments for activities that could have adverse transboundary effects; translation of the Rio/UNCED principles on forestry into a binding treaty with target dates, costs, funding sources and a democratic, transparent, equitable financial mechanism to support it; development of a treaty or treaties on liability and compensation for environmental damage; and development of a protocol to the Biodiversity Convention covering safety requirements for the use of genetically engineered organisms. These are the salient issues at this writing; as time goes on others will doubtless arise, and scientists, economists, the medical profession, and NGOs should press the international community for action rather than waiting for an environmental disaster to encourage the media to take the lead.

In addition to the issues sketched above, changing consumption and production patterns will require a mix of theory and policy. The problem of how to change consumption patterns was an issue first placed on an agenda for multilateral negotiation at the UNCED in 1992. Policies and measures to change production and consumption patterns should be predictable and support sustainable development. Once the true life cycle cost of an item is determined, its price can be established, and a new kind of market force will drive the consumption decisions of nations (especially the developed nations), business, industry, and individual households. Under these guidelines, prevention of pollution, suitably encouraged, would result in cost reduction. Consumption and production patterns can be changed, and to some extent have changed already, through instruments encouraging energy conservation; use of renewable energy sources; use of public transport; waste reduction, recycling and reuse; less packaging; use of "green" products and their development; conservation of water; and reduction of environmentally harmful substances in consumer products. A better understanding of the interrelationship among consumption patterns, production structures and techniques, economic growth, employment, population dynamics, and environmental stress must be achieved, together with the establishment of studies of the effects of present consumption and production patterns and an assessment of their sustainability and their effects on the world economy.

Funding the Global Environment

It is now generally accepted that the delicate planetary balances are threatened; some, such as the ozone layer, the CO_2 balance, and fresh-water reserves, have already been damaged. An effective response must include a sense of shared responsibility to the future, participated in by the developing as well as the developed countries. Laying aside questions of responsibility for the past, North and South must work out equitable means of reducing or even reversing the damages. This means that the wealthier and poorer nations—to use the old meaning of the terms—must exchange binding commitments, the latter to implement agreed solutions, the former to make available technology and funds to enable them to do so.

Complicating this arrangement is the fact that advanced technologies, particularly new and emerging technologies, belong to industry and not governments. Industrial conglomerates have spent considerable amounts to develop them and go to great lengths to protect their investment with patents, proprietary rights, royalty agreements, and so on. Until their stockholders can be convinced that the long-term health of the planet is worth more to them than this year's return on their investment, contributions from this source cannot be counted on.

Until recently voluntary contributions have paid for a large part of social and economic development, but this process has been losing momentum. Donor nations, which originally pledged an average of 0.7 percent of GNP for such assistance, have actually contributed much less, from 0.35 percent in 1992 down to 0.27 percent in 1995. So, again, the trend indicates that donors cannot be counted on.

The long-term nature and magnitude of the problem is such that the resources needed will be greater than in the past, sustainable into the indefinite future, predictable, and assured. Economists agree that air and water pollution on a global scale damages the global commons, and costly defensive measures whose benefits can only be assessed on an international scale cannot be carried out on the basis of commercial loans or conventional assistance programs. New mechanisms must be devised. The World Bank and other members of the capital markets cannot provide enough funds; taxation is not the answer, particularly in the traditional donor countries; and it is likely that in the near term many of the resources that have gone to the developing world will be diverted to eastern Europe.

In the past twenty-five years only one of the important environmental agreements—the Montreal Protocol—has been designed with its own funding mechanism. In spite of studies by specialists in international financing, no final blueprint has emerged, and interest in the topic is sporadic at best. The developing countries still have only two choices: to borrow (unlikely, with their $1.3 trillion of foreign debt) or to ask yet again for aid, which is unlikely to be forthcoming. However, with the growing understanding of the extent to which the environment has deteriorated an awareness is developing of how such development ingredients as agriculture, fisheries, energy, transportation, and health are affected. Development needs can be expressed in environmental terms; this new language is likely to be much more widely accepted than yet another appeal for foreign aid. The economic rationale for increasing the resources available for international needs is that the economic and environmental problems of one country or region can no longer be viewed in isolation; costly measures to safeguard or repair the global commons, whose benefits will be international in scope, cannot be carried out with traditional means. New and innovative methods must be sought.

In the search for such mechanisms, there must be a free and open partnership between the developed and developing countries, an acceptance of joint responsibility for pollution and thus for the costs of its reversal. Nor may the funds already earmarked for development purposes be diverted; any new funding must be additional. It must be predictable, continuous, and increasingly assured, and flexibly devised to meet future conditions. Not only are we seeking a considerable increase in financial resources, we also seek to reduce damaging consumption patterns as much as possible. This points to the selective use of incentive instruments such as user's fees, consumption taxes, and the like.

Any new funding mechanism should be administratively feasible, equitable in terms of burden sharing, and simple to collect. User's fees seem to meet these three criteria. They could be combined with a system of credit offsets to reward countries for special efforts on behalf of the environment. As an example of how user's fees would work, let us consider the ozone depletion problem. All of us are to some extent destroying ozone through air conditioning in cars, houses, and offices, through refrigerators and foams, in the manufacture of televisions and video equipment, and in the use of halons in fire extinguishers. Just as we pay for the water we use, why not consider clean air a similar resource? When we diminish that quality that makes air valuable, its cleanliness, we would pay a user's fee. These fees would be contributed to an international fund and disbursed to make patents or technology available to a developing country.

In the case of climate change, were fees to be imposed for CO_2 production, special subsidies could ensure the maintenance of existing forests. Although considerable emphasis has been placed on reforestation, maintaining present forest communities, with their rich genetic heritage, should be given more importance. In a feature similar to farm subsidies, countries could be rewarded for foregoing exploitation of their forest resources in the interest of ecological stability. Credit could be extended to those who maintain their forests, reduced to those who cut them, and increased to those who replant them. Many details need to be worked out and collection and disbursement mechanisms identified; an agreement in principle would be needed to begin the process. These arrangements do not depend on technical feasibility but on political will.

Even if funds were made available, it is asked, what is the guarantee that patents and know-how would be forthcoming? Unless issues of mistrust, legal requirements, and financial guarantees are resolved, the most carefully constructed environmental treaties and action plans may fall apart.

As early as 1970, the Committee for Development Planning of the United Nations considered a proposal to tax certain consumer goods, a sort of polluter-pays tax that would have been levied on private aircraft, pleasure boats, automobiles, television sets, refrigerators, air conditioners, washing machines, and other items that, although common in the developed world, are luxuries in the developing nations. Each country would collect its own tax revenues under the proposal. In the same period, the Brandt Commission recommended establishment of a World Development Fund, a supplement to existing funds, to channel resources mobilized through automatic means. However, neither plan featured the global-partnership approach that would characterize the user's fee system, involving full participation in decision making by every country on an equal footing.

What is called for is an investment in the future. Nations of the world may not have a common past, but they have a stake in a common future. A new system of priorities based on global partnership must be established in which everyone pays a fair share into a common pool of resources to combat environmental danger. This aspect should make the concept of user's fees acceptable to the public.

Technology Regulation and Transfer

Today's environmental problems are largely a legacy of yesterday's technology, but that legacy's forward-looking aspect promises solutions thanks to an ongoing revolution in such fields as biotechnology, information systems, telecommunications, miniaturization, and new materials. This revolution can spur the greening of industry by allowing firms to attain the same output while using less energy and fewer raw materials, producing less pollution and waste. Technical innovations in pollution monitoring and control have already matured into an environmental industry.

While new technologies sweep through industrialized nations, the rest of the world has neither the capital nor the scientific, social, and legal infrastructure to take full advantage of them. The international community must find ways to speed the flow of advanced technology to developing countries. To do this, government, industry, and the scientific community must evolve new policies, adding technological cooperation to existing business charters, creating transitional commercial networks to spread technologies, and establishing intermediaries who would bring technologies together with potential users. Funding for this effort could be provided by foundations, development assistance agencies, and business consortia. The "build, operate, and transfer" concept could also be applied to the new technologies.

The new science of biotechnology offers an example of both the promise of the new technologies and the dangers of their unregulated growth. The field is made up of two groups: genetic engineering, based on advances in molecular biology, biochemistry, and genetics; and cellular engineering, based on the earlier practices of tissue culture. Genetic engineering, which theoretically allows any gene to be moved from one organism to another, although it originated in universities and small firms, is now almost entirely under the control of transnational enterprises. Need for regulation of this industry was seen as long ago as 1975, when part of the scientific community led by Paul Berg, a molecular biologist from Berkeley, expressed its concerns about the new technology. The move for regulation faded away as many of the scientists became involved in commercial applications. The recent success of a sheep-cloning experiment in Scotland revived concerns in the scientific community, focusing on ethical issues. Sustaining analysis of the social impact of the new technologies has become the responsibility of individual scientists and activists. The most persistent fear has been of the consequences of an accidental or deliberate release of self-propagating genetically engineered organisms into the biosphere. This was exemplified in 1983 with the "ice minus" incident.

In attempting to improve the frost tolerance of crops, biotechnologists isolated a gene that triggers ice nucleation in plant cells. They deleted it from the bacterium *Pseudomonous syringae*, and in 1983 Steven Lindow and his funding organization, Advanced Genetic Sciences, received permission from the U.S. National Institutes of Health to field test the process by spraying it on a crop. The theory was that the bacterium would displace the naturally occurring ice-forming mechanism, lowering the freezing point of the plants. However, a group of citizens and environmental groups filed suit against the NIH, charging that it had not adequately assessed the project's potential environmental risks. They listed the dramatic possibility that the bacteria might be swept into the upper atmosphere, disrupting the formation of ice crystals and ultimately affecting local weather patterns and possibly altering the global climate. Eminent scientists pointed to the ecological hazards posed by deliberate release of microorganisms because they reproduce rapidly and their interrelations with higher plant forms are unknown.

The first hormone of the new biological generation is bovine growth hormone, BST. Natural BST regulates muscle formation and growth in young cattle and milk production in adult cows. Genetically engineered BST is produced not by cows but by engineered bacteria. Administered to cows daily, it increases milk yields by 7–14 percent; however this leads to severe deterioration of the health of the cow and to overproduction of milk in areas where milk surpluses are already driving small dairy farmers out of business. BST has been banned in Denmark, Sweden, and Norway, and the European Parliament supported a resolution calling for a worldwide ban on BST.

Although bans and regulations in the North seem to be a hopeful sign, in fact genetically engineered products are increasingly being tested or distributed in the South in order to avoid regulation and public control. Ignorance about the ecological and health impacts of the new technologies far outweighs the knowledge needed to produce them. The unintended results of applications of new technologies often produce more damage than benefit: it took 200 years of dependence on fossil fuels to recognize the damage caused to our climate; DDT, celebrated as a tool to ensure public health, which won a Nobel Prize for its inventor, carried very high costs, and it and other toxic pesticides have been banned in the industrialized countries; Union Carbide, in setting up chemical plants in India, said proudly, "We have a hand in India's future," but in December 1984, a deadly gas leaked from its pesticide plant in Bhopal and killed 3,000 people.

Hazardous substances and processes have been manufactured faster than public controls and regulations can evolve. Testing in even such widespread and familiar areas as fossil fuel technologies and chemical engineering is still in its infancy; in the genetic engineering field such tests have yet to be conceived. How genetically modified life forms interact with other organisms is uncharted territory. Furthermore, these products, once developed, cannot be removed from the market. Here more than in any other area, lack of knowledge cannot be construed as safety. The only wise strategy is to use restraint and caution.

Technology transfer must be negotiated within the framework of an assessment of its ecological, social, and economic impact, so that socially desirable transfer can take place and undesirable and hazardous transfer prevented. Criteria must be devised to distinguish between the two, and the tendency avoided to treat technology as an end in itself, not as a means to reach a goal that may be better approached more conservatively.

Probably the most serious impact of biotechnology will be the displacement of some agricultural export commodities from the Third World, with its related impacts on the national economy. Many high-value products derived from plants and used for pharmaceuticals, dyes, flavorings, and fragrances are already being displaced by engineered substitutes. This movement will be felt most by countries that were earlier made dependent on exports of natural products, Africa in particular, where some economies rely on single crops for most of their export earnings.

When factories close in the North, compensation may be given to workers; when crops first introduced by global agribusiness are displaced by technologies developed by agribusiness, the peasant and agricultural worker are left to fend for themselves. An agenda for compensation must be developed based on historical justice and placed on the global negotiating table before the new biotechnologies are fully deployed.

Building International Environmental Accords

In the course of the many meetings that resulted in the six international agreements described in previous chapters, a number of factors emerged. It was obvious throughout the negotiations that governments kept returning to the same issues over and over again: entry into force, how often the Conference of the Parties should meet and the limits of its responsibility, the composition of the Secretariat, mechanisms for settling disputes, and withdrawal, among others. An attempt to develop standard language to cover such issues failed, as the international lawyers were very reluctant to do so. Negotiations for environmental treaties could be considerably shortened, saving money and effort for more urgent matters, if these routine issues could be standardized. We propose this as a timely subject for scholars of international law.

There has been a tendency for the negotiations to be dominated by the most powerful states, and in some cases not all groups holding stakes in the outcome were adequately represented. There was often a focus on feared losses due to environmental protection, rather than the gains to be expected through a sharing of economic and ecological benefits.

Regarding the makeup of the delegations, it was a common feature that earlier meetings were attended by minor functionaries of the various government ministries, and it was not until the final stages of negotiations that higher-ranking representatives attended. This meant that they were often uninformed of the proceedings of the previous sections. This lack of communication was particularly notable among the developing countries and was exacerbated by the continual change of experts from these countries, which required going over the same issues more than once in some cases. Generally the delegations were led by representatives of ministries of foreign affairs or the environment, or their equivalent. Almost all the ministers who attended the plenipotentiary conferences, however, were ministers of the environment. Delegations also usually included representatives from their permanent missions to the UN, mainly from Geneva offices, because most of the meetings were held in Geneva.

The OECD and the European Community played a very active role in the negotiations, but neither was treated as a government. Although the EC was given the right to sign and ratify the various conventions and become a party to them, it was seated during the negotiations, not in its alphabetical place among governments, but among the intergovernmental organizations.

The OECD was particularly active in negotiating the Basel Convention, having worked for several years on its own regional convention on transboundary movement of hazardous waste. This convention was not adopted, however, because only the OECD members had been involved in its development.

Several NGOs were active participants in the negotiations of all the treaties we have described. Most of them had a lively awareness of the problems and were effective in keeping the general public informed. They kept up an active liaison with several developing countries' governments so as to have them present their views when their interests coincided. The most active of these groups were the Natural Resources Defense Council (NRDC); the International Organization of Consumers' Unions (IOCU); Greenpeace International; the Entente Europeen pour l'Environnement (EEE); and the International Environment Liaison Centre (IELC).

Industry and business were highly visible at these negotiations. The subjects of the various instruments touched very closely on their activities, and their input was valuable. Their common view was that although they appreciated the positive role of the NGOs in trying to solve some of the problems dealt with, in some cases it had been the NGOs that triggered the problems, especially that of the transboundary movement of hazardous waste. They felt that by publicizing the issue the NGOs had raised public resistance to disposal in the developed countries, even when the disposal sites were environmentally safe, and thus precipitated the search for overseas disposal. Prominent during the negotiations were the Chemical Manufacturers' Association (CMA), the Conseil Europeen des Federations de l'Industrie Chimique (CEFIC), the International Chamber of Commerce (ICC), Imperial Chemical Industries (ICI), and DuPont, among others.

Other participants attended as observers, with their chairpersons invited to address the meetings of the working groups. However, on some occasions, when negotiations were difficult and tempers flared, the observers were excluded. This happened most often during the negotiations for the Montreal Protocol.

Epilogue

The international community has embarked on a new experiment. It is entering a new dimension, and a society is taking shape whose commerce is geared to the international marketplace and rests upon the gifts of the global environment. But if it is to have any meaning, it must embody a common ideal and a cohesive philosophy; it must be defined by the ethical and moral principles it claims to embrace. It must be strong enough to propel widely divergent national histories, suspicions, and aspirations toward a common future. We ask, what kind of common future do we want? Some twenty years ago, former Canadian Prime Minister Pierre Trudeau called for a "just society." Today the call is for a "just global society."

There are many areas where global justice is painfully absent and international responsibility must be brought to bear. They include poverty, famine, disease, abuse of human rights, and misuse and abuse of natural resources. In *Facing Mount Kenya*, Jomo Kenyatta wrote:

> A man is the owner of his land. . . . But insofar as there are other people of his own flesh and blood who depend on that land for their daily bread, he is not the owner, but the partner, or at the most a trustee for the others. Since the land is held in trust for the unborn as well as for the living, and since it represents his partnership in the common life of generations, he will not lightly take it upon himself to dispose of it.

Our world would do well to follow Kenyatta's description of the special kinship between man and his land. National boundaries should not obscure the fact that all people in the North and South, the East and West, are united; we are all partners and trustees for the unborn. This is, unfortunately, an aspiration quite distant from today's reality. Humanity has disposed of nature's gifts both lightly and foolishly.

While the engines of environmental destruction take an unimaginable toll, our knowledge of our biological inventory has merely skimmed the surface of nature's miracle. Two decades after humans walked the surface of the moon, we know very little about the secrets of our Earth.

The plight of the world's tropical forests in the Amazon basin, western Africa, and Southeast Asia has attracted worldwide attention: as the twentieth century began, nearly all tropical forests that had existed for twenty centuries were intact, but in less than 100 years half of them have disappeared. Today, approximately 100 acres are destroyed each minute; at this rate, all tropical forests will disappear in the next century, and the home of nearly half our plant's known species will be destroyed, including those that may hold the secrets to curing cancer, AIDS, heart disease, and other diseases.

The global population, currently at just over 5.1 billion, has doubled since 1954 and is predicted to double again by 2050. Meanwhile soil is being lost at a record rate. Consider these statistics: In two and a half days more than a million babies are born and 50,000 hectares of topsoil lost through erosion. Although economic growth has brought great benefits to mankind, they are being hoarded by too few. Less than a quarter of the world's population enjoys nearly three-quarters of its wealth. If our world is to survive, economic growth must increase, the poor must be allowed to share the world's wealth, and the rich must curb their demands and practice conservation, recycling, resource efficiency, and economic equity.

Signs of Hope

There are very clear, strong, and durable signs of hope. The world is learning that economic sustainability relies on environmental sustainability. Changes in agricultural practices enable small farmers to be less vulnerable to changes in the volatile commodity markets and more focused on the needs of their local communities. Progress continues in many developing countries toward reforming land tenure, reducing chemical use, encouraging biological fertilization, and improving pest management and irrigation methods. Experiments in hybrid seed pools enable natural forests to coexist with agricultural land. Major efforts are also underway to improve energy efficiency, not only in the affluent societies, but also among the poor, who are using affordable, fuel efficient stoves, and also through more appropriate energy planning.

Environmentally sound, sustainable development is a realistic plan for the future. It has been widely embraced. It can and will work. But it depends on the unwavering support not only of today's leaders, but also of the young, tomorrow's leaders, to ensure that our planet's resources are protected, its future secured.

International action to safeguard our environment is taking shape, and a strong consensus has emerged geared to meaningful and realistic action. It recognizes the integrity of global environmental justice, whereby common efforts are urgently needed to protect our Earth, to assist those most at risk and least responsible for environmental deterioration: the world's poorest. The easing of East-West tensions presents a golden opportunity to further the goals of environmental justice.

Despite the enormous problems we have outlined, we wish to conclude with a message of hope. We are more optimistic about our environmental future today than at any time during the past two and a half decades of service to the international community. The reality of a global society is finally being seen; the promise of a wider perception of justice is emerging. Governments, industry, and the public, alerted by a long series of environmental catastrophes, now recognize that we all share the only planet we have. The future on which we have embarked is the only future we have. We have nowhere else to go.

Glossary of Acronyms

ASEAN	Association of Southeast Asian Nations
CGIAR	Consultative Group on International Agricultural Research
CITES	Convention on the International Trade in Endangered Species
CMS	Convention on Migratory Species
DAC	Development Aid Committee
EC	European Community (until 1994; now European Union)
ECE	United Nations Economic Commission for Europe
EEC	European Economic Community
EU	European Union
FAO	United Nations Food and Agriculture Organization
GATT	General Agreement on Tariffs and Trade
GEF	Global Environment Facility
IBPGR	International Board for Plant Genetic Resources
ILO	International Labour Organisation
IMO	International Monetary Organization
IPCC	Intergovernmental Panel on Climate Change
IPCS	International Programme on Chemical Safety
IRPTC	International Register of Potentially Toxic Chemicals
IUCN	International Union for the Conservation of Nature (now known as the International Conservation Union)
OECD	Organization for Economic Cooperation and Development
RAMSAR	Convention on Wetlands of International Importance Especially as Waterfowl Habitat

UNCED	United Nations Conference on Environment and Development
UNCTAD	United Nations Conference on Trade and Development
UNDP	United Nations Development Program
UNESCO	United Nations Educational, Scientific and Cultural Organization
WFP	United Nations World Food Program
WIPO	World Intellectual Property Organization
WMO	World Meteorological Organization
WWF	World Wide Fund for Nature (formerly World Wildlife Fund)

Suggested Readings

Albin, Cecilia. 1992. "Fairness issues in negotiation: Structure, process, procedures and out-come." Working paper No. WP-92-88, International Institute for Applied Systems Analysis, Laxenburg, Austria.

Antrim, Lance, and P. Chased. 1992. "The UNCED negotiating process." *Ocean and Coastal Management* 18(1).

Barratt-Brown, Elizabeth P. 1991. "Building a monitoring and compliance regime under the Montreal Protocol." *Yale Journal of International Law* 16(2).

Bates, John H., and Charles Benson. 1992. "Marine environment law." Book review by Peter Wetterstein. In *Yearbook of Environmental Law,* vol. 3. London: Graham & Trotman.

Benedick, Richard. 1991. *Ozone Diplomacy: New Directions in Safeguarding the Planet.* Cambridge, MA: Harvard University Press.

Birnie, Patricia W., and Alan E. Boyle. 1993. "International law and the environment." Book review by Stephen Mc Caffrey. In *Yearbook of Environmental Law,* vol. 4. London: Graham & Trotman.

Biswas, A. et al. 1983. *Long-Distance Water Transfer.* Dublin: Tycooly International.

Bodansky, Daniel. 1992. "Managing climate change." In *Yearbook of Environmental Law,* vol. 3. London: Graham & Trotman.

Boehmer-Christiansen, S. A. 1992. "How much science does environmental performance really need?" In *Achieving Environmental Goals*, ed. E. Lykke. London: Belhaven.

Bolin, Bert, Bo Doos Warrick, and J. Jager, eds. 1986. *SCOPE 29: The Greenhouse Effect—Climatic Change and Ecosystems.* Chichester, UK: Wiley and Sons.

Brown-Weiss, Edith, Daniel B. Magraw, and Paul C. Szasz. 1992. "International environment law: Basic instruments and documents." In *Yearbook of Environmental Law,* vol. 3. London: Graham & Trotman.

Burhenne-Guilmin, Françoise, and Susan Casey-Lefkowitz. 1992. "The new law of biodiversity." In *Yearbook of Environmental Law,* vol. 3. London: Graham & Trotman.

Burns, P. 1988. "Hazardous waste management—The way forward." *Journal of Institute of Water & Environmental Management* 2:285.

Carnevale, Peter, and Dean Pruitt. 1992. "Negotiation and mediation." *Annual Review of Psychology* 43.

Carroll, John E., ed. 1988. *International Environmental Diplomacy*. Cambridge, UK: Cambridge University Press.

Carson, Rachel. 1962. *Silent Spring*. Boston: Houghton-Mifflin.

Chayes, Abram, and Antonia Chayes. 1991. "Compliance without enforcement: State behavior under regulatory treaties." *Negotiation Journal*.

Cross, John. 1978. "Negotiation as a learning process." In *The Negotiation Process, Theories and Applications*, ed. I. William Zartman. Beverly Hills, CA: Sage Publications.

Deegan, J. 1987. "Looking Back at Love Canal." *Environmental Science and Technology* 21:328.

Ehrlich, P., and A. Ehrlich. 1982. *Extinction: The Causes and Consequence of the Disappearance of Species*. London: Victor Gollancz.

El-Hinnawi, E. 1991. *Sustainable Agriculture and Rural Development in the Near East*. Regional Document No. 4, FAO/Netherlands. Rome: FAO, Conference on Agriculture & Environment.

El-Hinnawi, E. 1981. *Environmental Impacts of Production and Use of Energy*. Dublin: Tycooly International.

El-Hinnawi, E., and A. Biswas. 1981. *Renewable Sources of Energy and Environment*. Dublin: Tycooly International.

Engelstad, O. P. 1984. "Crop nutrition technology." In *Future Agricultural Technology and Resource Conservation*, ed. B. C. English et al. Ames: Iowa State University Press.

EPA. 1988. *Environmental Progress and Challenges: EPA's Update*. EPA No. 230-07-88-033. Washington, DC: United States Environmental Protection Agency.

Falkenmark, M. 1986. "Fresh waters as a factor in strategic policy and action." In *Global Resources and International Conflict*, ed. A. H. Westing. Oxford: Oxford University Press.

FAO. 1991. *The State of Food and Agriculture—1990*. Rome: United Nations Food and Agriculture Organization.

FAO. 1989. FAO Conference, 25th session, resolution 6/89.

FAO. 1988. *Country Tables*. Rome: United Nations Food and Agriculture Organization.

FAO. 1987. FAO Conference, 24th session, resolution 5/87.

FAO. 1985. FAO Conference, 23rd session, resolution 10/85.

Farman, J. C. et al. 1985. "Large losses of total ozone in Antarctica reveal CLO_x-NO_x interaction." *Nature* 315:207.

Francioni, Francesco, and Tullio Scovazzi, eds. 1992. *International Responsibility for Environmental Harm.* Book review by Stephen McCaffrey. In *Yearbook of International Environmental Law,* vol. 3. London: Graham & Trotman.

Fullick, A., and P. Fullick. 1991. "Biological pest control." *New Scientist,* Issues in Science 43(9).

Gehring, Thomas. 1990. "International environmental regime and dynamic sectoral legal systems." *Yearbook of International Environmental Law,* vol. 1. London: Graham & Trotman.

GEMS/WHO. 1989. *Global Freshwater Quality: A first assessment.* Oxford: Blackwell.

General Agreement on Trade and Tariffs (GATT). 1982. Report of Ministerial Meeting, November. Geneva, Switzerland.

GESAMP. 1990. *The State of the Marine Environment.* Regional Seas Reports and Studies No. 115. Nairobi, Kenya: UNEP.

Glotfelty, D. E., et al. 1987. "Pesticides in fog." *Nature* 325:602.

Haas, Peter. 1990. *Saving the Mediterranean: The Politics of International Environmental Cooperation.* New York: Columbia University Press.

Handl, Gunther. 1990. "Environmental security and global change: The challenge to international law." *Yearbook of International Environmental Law,* vol. 1. London: Graham & Trotman.

Herter, Christian, Jr., and Jill Binder. 1993. *The Role of the Secretariat in Multilateral Negotiation: The Case of Maurice Strong and the 1972 U.N. Conference on the Human Environment.* From the Case Study Series of the Project on Multilateral Negotiation. Washington, DC: American Academy of Diplomacy and the Paul H. Nitze School of Advanced International Studies, Johns Hopkins University.

Hohmann, Harald, ed. 1992. *Basic Documents of International Environmental Law,* vols. 1–3. London: Graham & Trotman.

Houghton, J. T., et al., eds. 1990. *Climate Changes: The IPCC Scientific Assessment.* Cambridge: Cambridge University Press.

Houghton, R. A. 1990. "The global effects of tropical deforestation." *Environmental Science and Technology* 24:414.

Hurrell, Andrew, and Benedict Kingsbury, eds. 1992. *The International Politics of the Environment.* Book review by Allen L. Springer in *Yearbook of Environmental Law,* 1993.

Huxley, William. 1964. *Essays of a Humanist.* London.

IPCC. 1990. *Report of the Intergovernmental Panel on Climate Change, WMO/UNEP.* Cambridge, UK: Cambridge University Press.

IPCC. 1990. Reports of Working Groups II & III, WMO/UNEP. Nairobi, Kenya: UNEP.

IUCN/UNEP/WWF. 1980. World Conservation Strategy. Gland, Switzerland: IUCN.

IUCN/UNEP/WWF. 1991. "Caring for the Earth: A strategy for sustainable living." Gland, Switzerland: IUCN.

Johnston, H. S. 1971. "Reduction of stratospheric ozone by nitrogen oxide catalysts from supersonic transport exhaust." *Science* 173:7311.

Kimball, Lee A. 1992. "Towards global environmental management: The institutional setting." *Yearbook of International Environmental Law,* vol. 3. London: Graham & Trotman.

Koh, Tommy T. B. 1992. "The Earth Summit's negotiating process: Some reflections on the art and science of negotiation." The Seventh Singapore Law Review Lecture, Faculty of Law, National University of Singapore.

Koskenniemi, Martti. 1992. "Breach of treaty or non-compliance? Reflections on the enforcement of the Montreal Protocol." *Yearbook of International Environmental Law*, vol. 3. London: Graham & Trotman.

Krasner, S., ed. 1983. *International Regimes*. Ithaca, NY: Cornell University Press.

Kremenyuk, Victor A. 1991a. "The emerging system of international negotiation." In *International Negotiation*, ed. Victor A. Kremenyuk. San Francisco: Jossey-Bass.

Kremenyuk, Victor, A., ed. 1991b. *International Negotiation: Analysis, Approaches, Issues*. San Francisco: Jossey-Bass.

Kressel, Kenneth, et al. 1989. *Mediation Research: The Process and Effectiveness of Third-Party Intervention*. San Francisco: Jossey-Bass.

La Riviere, J. W. 1989. "Threats to the world's water." *Scientific American* 261:48.

Lang, Winfried. 1989. "Multilateral negotiations: The role of presiding officers." In *Processes of International Negotiations*, ed. F. Mautner-Markhof. Boulder, CO: Westview Press.

Lang, Winfried. 1992. "Diplomacy and international environmental law-making: Some observations." *Yearbook of International Environmental Law* 3:113. London: Graham & Trotman.

Langkawi Declaration on Environment. 1989. "Development and international economic cooperation: Environment." New York: United Nations A/44/673.

Lean, G., et al. 1990. *Atlas of the Environment*. London: Arrow Books.

Marsh, George Perkins. 1864. "The Earth as modified by human action." *Man and Native*. London.

McDonald, John W. 1990. "Global environmental negotiations: The 1972 Stockholm Conference and lessons for the future." Occasional Paper #2, American Academy of Diplomacy.

McNeely, J. A., et al. 1990. "Conserving the World's Biological Diversity." Gland, Switzerland: IUCN.

Meadows, Denis, et al. 1972. *The Limits to Growth*. Washington, DC: Potomac Associates.

Molina, Mario J., and F. S. Rowland. 1974. "Stratospheric sink for chlorofluoromethane." *Nature* 249:810.

Moomaw, William R. 1994. "Protecting the ozone layer: A revolutionary approach to evolutionary treaties." In *Transnational Environmental Law and its Impact on Corporate Behaviour: A Symposium on the Practical Impact of Environmental Laws and International Institutions on Global Business Development*, ed. Eric J. Urbaniy, Conrad P. Rubin, and Monica Katzman, 329–338. Irvington, NY: Transnational Juris Publications.

Moore, J. N., and S. N. Luoma. 1990. "Hazardous wastes from large-scale metal extraction." *Environmental Science and Technology* 24:1278.

NASA. 1979a. *Stratospheric Ozone Depletion by Halocarbons*. Washington, DC: National Academy Press.

NASA. 1979b. *The Stratosphere: Present and Their Effects on Stratospheric Ozone*. Pollution Paper no. 15, UK Department of Environment, Her Majesty's Stationery Office, London.

OECD. 1991. *The State of the Environment—1991*. Paris: OECD.

OECD. 1984. Recommendation 84 (c) 37.

Parikh, Jyoti K. 1992. "IPCC Response Strategies: Unfair to Poor?" *Nature* (December).

Parry, Martin L., et al. 1990. "The potential impact of climate change on agriculture and forestry." Draft report, Intergovernmental Panel on Climate Change.

Parther, M. J., et al. 1984. "Reductions in ozone at high concentrations of stratospheric halogens." *Nature* 312:227.

Pimentel, D., and L. Levitan. 1986. "Pesticides: Amounts applied and amounts reaching presets." *BioScience* 36:86.

Porter, Gareth, and Janet Brown. 1991. *Global Environment Politics*. Boulder, CO: Westview Press.

Prescott-Allen, C., and R. Prescott-Allen. 1986. *The First Resource: Wild Species in the North American Economy*. New Haven, CT: Yale University Press.

Pruitt, Dean. 1981. *Negotiation Behavior*. New York: Academic Press.

Raiffa, Howard. 1982. *The Art and Science of Negotiation*. Cambridge, MA: Belknap Press/Harvard University Press.

Raven, P. H. 1988. "Our diminishing tropical forests." In *Biodiversity*, ed. E. O. Wilson. Washington, DC: National Academy Press.

Repetto, R. 1986. *Paying the Price: Pesticide Subsidies in Developing Countries*. Research Report no. 2. Washington, DC: World Resources Institute.

Repetto, R., et al. 1989. *Wasting Assets: Natural Resources in the National Income Accounts*. Washington, DC: World Resources Institute.

Rittberger, Volker. 1983. "Global conference diplomacy and international policy-making: The case of UN-sponsored world conferences." *European Journal of Political Research* 11.

Rothman, Jay. 1992. *From Confrontation to Cooperation: Resolving Ethnic and Regional Conflict.* Newbury Park, CA: Sage Publications.

Rowland, F. S. 1991. "Stratospheric Ozone in the 21st Century." *Environmental Science and Technology* 25:622.

Rowland, F. S. 1990. "Stratospheric ozone depletion by chlorofluorocarbons." *Ambio* 19:281.

Rowland, F. S. 1987. "Can we close the ozone hole?" *Technology Review* (August/September):51.

Rowland, F. S., and Molina, M. J. 1975. "Chlorofluoromethanes in the environment." *Revue of Geophysics and Space Physics* 13:1.

Rubin, Jeffrey, and Bert Brown. 1975. *The Social Psychology of Bargaining and Negotiation.* New York: Academic Press.

Rummel-Bulska, Iwona. 1994. *The Basel Convention: A Global Approach for the Management of Hazardous Wastes.* Bonn, Germany: Environmental Policy and Law.

Rummel-Bulska, Iwona. 1991a. *Environmental Law in UNEP.* UNEP Environmental Law Library, no. 1. Nairobi, Kenya: UNEP.

Rummel-Bulska, Iwona. 1991b. *Selected Multilateral Treaties in the Field of the Environment.* Vol. 2, edited by Iwona Rummel-Bulska and Seth Osafo Grotius. Cambridge: Publications Limited.

Rummel-Bulska, Iwona. 1991c. *Activities of UNEP in the Field of Environmental Law in* 1991. UNEP Environmental Law Library, no. 4. Nairobi, Kenya: UNEP.

Rummel-Bulska, Iwona. 1984. "The UNEP African Inland Water programme." In *Year Book of the AAA: Natural Resources in International Law.* The Hague, the Netherlands: The Hague Academy of International Law.

Rummel-Bulska, Iwona. 1981. *The Uses of Inland Waters for Non-Navigational Purposes, in the Light of International Law.* Warsaw: Polish Academy of Science.

Rummel-Bulska, Iwona. 1980. "The responsibility of states for damages caused by use of air and water." In *Responsibility of States in International Law*, ed. I. Rummel-Bulska. Warsaw: Polish Institute of International Law.

Rummel-Bulska, Iwona. 1972. "The protection of air in the national legislation of some countries in the world." *Journal of International Affairs.*

Rummel-Bulska, Iwona, and K. Kummer. 1990. *The Basel Convention.* UNEP Environmental Law Library, no. 2. Nairobi, Kenya: UNEP.

Rummel-Bulska, Iwona, and A. Tolentino. 1991. "Legal and Institutional Arrangements for Environmental Protection and Sustainable Development in Developing Countries." UNEP Environmental Law Library, no. 3. Nairobi, Kenya: UNEP.

Sand, Peter H. 1992a. "The Effectiveness of International Environmental Agreements." Book review by Andronico O. Adede. In *Yearbook of International Environmental Law*, vol. 3. London: Graham & Trotman.

Sand, Peter H. 1992b. "Lessons Learned in Global Environmental Governance." Book review by Martti Koskenniemi. In *Yearbook of International Environmental Law*, vol. 3. London: Graham & Trotman.

Sand, Peter H. 1992c. "UNCED and the Development of International Environment Law." *Yearbook of International Environmental Law*, vol. 3. London: Graham & Trotman.

Sand, Peter H. 1990. *Lessons in Global Environmental Governance.* Washington, DC: World Resources Institute.

Schally, Hugo M. Forests. 1993. "Toward an International Legal Regime?" *Yearbook of International Environmental Law,* vol. 4. London: Graham & Trotman.

Schneider, C. 1988. "Hazardous waste: The bottom line is prevention." *Issues in Science and Technology* 4:75.

Sjostedt, Gunnar, ed. 1993. *International Environmental Negotiations.* Newbury Park, CA: Sage Publications.

Sjostedt, Gunnar, and Bertram I. Spector. 1993. "Conclusions." In *International Environmental Negotiation,* ed. G. Sjostedt. Newbury Park, CA: Sage Publications.

SMIC. 1971. "Inadvertent climate modification: Report of the study of man's impact on climate." Symposium hosted by the Royal Swedish Academy of Sciences and the Royal Swedish Academy of Engineering Sciences. Cambridge, MA: MIT Press.

South Commission. 1990. *The Challenge to the South.* Oxford: Oxford University Press.

Spector, Bertram. 1993a. "Decision analysis: Evaluating multilateral negotiation processes." In *International Multilateral Negotiation,* ed. I. William Zartman. San Francisco: Jossey-Bass.

Spector, Bertram I. 1993b. "Decision analysis for practical negotiation application." *Theory and Decision* 34.

Spector, Bertram, ed. 1993c. International Negotiation: Theory, Methods, and Support Systems. *Theory and Decision* (special issue) 34(3).

Spector, Bertram I. 1992. *International Environmental Negotiation; Insight for Practice.* Executive Report. Working paper, IIASA, Laxenburg, Austria.

Spector, Bertram I., and Anna R. Korula. 1992. *Facilitative Mediation in International Disputes: From Research to Practical Application.* Working paper, IIASA, Laxenburg, Austria.

Spector, Bertram I., and Anna R. Korula. 1993. "Problems of Ratifying International Environmental Agreements: Overcoming Initial Obstacles in the Post-Agreement Negotiation Process."

Springer, Allen L. 1983. *The International Law of Pollution.* London: Quorum Books.

Stein, J., ed. 1989. *Getting to the Table.* Baltimore, MD: Johns Hopkins University Press.

Supanich, Gary. 1992. "The legal basis of intergenerational responsibility: An alternative view—The sense of intergenerational identity." *Yearbook of International Environmental Law*, vol. 3. London: Graham & Trotman.

Susskind, Lawrence E. 1994. *Environmental Diplomacy: Negotiating More Effective Global Agreements.* New York: Oxford University Press.

Szell, Patrick. 1993. "Negotiations on the ozone layer." In *International Environmental Negotiation,* ed. Gunnar Sjostedt. Newbury Park, CA: Sage Publications.

Szenes, E., and N. Zoltai. 1988. "The Hungarian experience in hazardous waste management." *Industry and Environment* 2:22.

Thacher, Peter. 1993. "The Mediterranean: A new approach to marine pollution." In *International Environmental Negotiation*, ed. Gunnar Sjostedt. Newbury Park, CA: Sage Publications.

Tolba, Mostafa K. 1992a. *A Committment to the Future.* London: UNEP.

Tolba, Mostafa K. 1992b. *Saving Our Planet.* London: Chapman and Hall.

Tolba, Mostafa K. 1988. *Evolving Environmental Perceptions from Stockholm to Nairobi.* Nairobi, Kenya: UNDP.

Tolba, Mostafa K. 1987a. "One Earth One Home." London: UNEP.

Tolba, Mostafa K. 1987b. *Sustainable Development, Constraints and Opportunities.* London: Butterworths.

Tolba, Mostafa K. 1983. "Earth Matters." London: UNEP.

Tolba, Mostafa K. 1982a. "Statement to the World Food Conference, Rome, November 1974." In *Development without Destruction*, ed. M. K. Tolba. Dublin: Tycooly International.

Tolba, Mostafa K., ed. 1982b. *Development without Destruction.* Dublin: Tycooly International.

Tolba, Mostafa K., and A. K. Biswas. 1991. *Earth and Us.* London: Butterworth-Heinemann.

Tolba, Mostafa K., et al. 1992. *The World Environment 1972–1992: Two Decades of Challenge.* London: Chapman and Hall.

Touval, Saadia, and I. W. Zartman. 1985. "Introduction: Mediation in theory." In *International Mediation in Theory and Practice*, ed. S. Touval and I. W. Zartman. Boulder, CO: Westview Press.

UN. 1990. *Achievements of the International Drinking Water Supply and Sanitation Decade, 1981–1990.* Report of the Secretary-General. New York: United Nations no. A/45/327.

UN. 1989. *Illegal Traffic in Toxic and Dangerous Products and Wastes.* Report of the Secretary-General. New York: United Nations, no. A/44/362.

UN. 1980. *Development and International Economic Co-operation.* New York: United Nations no. A/35/592/Add.1.

UN. 1974. *The Cocoyoc Declaration Adopted by the Participants in the UNEP/UNCTAD Symposium on Patterns of Resource Use, Environment and Development Strategies held at Cocoyoc, Mexico from 8 to 12 October 1974.* New York: United Nations no. A/C.2/191.

UN. 1972. The United Nations Conference on the Human Environment, Stockholm from 5 to 12 June 1972. New York: United Nations A/C.2.

UN. 1971. *Development and Environment.* Reports submitted by a panel of experts convened by the Secretary-General of the United Nations Conference on the Human Environment, Founex, Switzerland, 4–12 June 1971. Stockholm: Kungl. boktryckeriet, PA Norstedt and Soner.

UN General Assembly. 1992. Resolution 37/137.

UN General Assembly. 1984. Resolution 39/229.

UNDP. 1991. *Human Development Report.* Oxford: Oxford University Press.

UNEP. 1994. Environmental Policy and Law 24/4 p. 147–148.

UNEP. 1991a. *Environmental Data Report.* 3rd ed. Oxford: Blackwell.

UNEP. 1991b. *The State of the Environment.* Nairobi, Kenya: UNEP.

UNEP. 1987. *Environmental Perspective to the Year 2000 and Beyond.* Geneva: UNEP.

UNEP. 1985. *Register of International Treaties and Other Agreements in the Field of Environment.* UNEP/GC/Info/11/Rev.1. Nairobi, Kenya: UNEP.

UNEP. 1984. *The State of the Environment, 1984.* Nairobi, Kenya: UNEP.

UNEP. 1982. *The World Environment, 1972–1982.* Dublin, Ireland: Tycooly International.

UNEP. 1980. "Choosing the Options: Alternative Lifestyles and Development Options." Nairobi, Kenya: UNEP.

Underdal, A. 1992. "The concept of regime effectiveness." *Cooperation and Conflict* 27(3).

Uriarte, F. A. 1989. "Hazardous waste management." In *ASEAN in Hazardous Waste Management,* ed. S. P. Maltrezon et al. London: Tycooly.

Van Dyke, John M., Durwood Zaelke, and Grant Hewison, eds. 1993. "Freedom of the seas in the twenty-first century: Ocean governance and environmental harmony." Book review by Alison Rieser. In *Yearbook of International Environmental Law,* vol. 4. London: Graham & Trotman.

Watson, R. J. 1988. "Current scientific understanding of stratospheric ozone." UNEP/Ozl. sc. 1/3. Nairobi, Kenya: UNEP.

Weiss, Edith Brown. 1988. *In Fairness to Future Generations: International Law, Common Patrimony and Intergenerational Equity.* Dobbs Ferry, NY: Transnational Publishers, Inc.

White, G. F. 1988a. "A century of change in world water management." In *Proceedings of the Centennial Symposium—Earth '88*. Washington, DC: National Geographic Society, 248.

White, G. F. 1988b. "The environmental effects of the high dam at Aswan." *Environment* 30:5.

Wilson, E. O. 1988. *Biodiversity*. Washington, DC: National Academy Press.

WMO. 1989. *Scientific Assessment of Stratospheric Ozone, 1989*. Global Ozone Research and Monitoring Project, Report no. 20. Geneva: WMO.

WMO. 1979. *Proceedings of the World Climate Conference*. Report no. 537. Geneva: WMO.

World Bank. 1991. *World Development Report, 1991: The Challenge of Development*. New York: Oxford University Press.

World Bank. 1990. *World Development Report, 1990*. Oxford: Oxford University Press.

World Bank. 1989. *World Development Report, 1989*. Oxford: Oxford University Press.

World Commission on Environment and Development. 1987. *Our Common Future*. Geneva: World Commission on Environment and Development.

World Conservation Monitoring Center. 1992. *Global Biodiversity—1992: Status of the Earth's Living Resources*. Cambridge, UK: WCMC.

World Health Organisation. 1990. *Public Health Impact of Pesticides Used in Agriculture*. Geneva: WHO.

World Resources Institute. 1990. *World Resources, 1990–1991*. New York: Oxford University Press.

World Resources Institute. 1987. *World Resources, 1987*. New York: Basic Books.

Yakowitz, H. 1989. "Global hazardous transfers." *Environmental Science and Technology* 23:510.

Zartman, I. William, ed. 1994. *International Multilateral Negotiations*. San Francisco: Jossey-Bass Publishers.

Zartman, I. William. 1993. "Lessons for analysis and practice." In *International Environmental Negotiation,* ed. Gunnar Sjostedt. Newbury Park, CA: Sage Publications.

Zartman, I. William. 1987. *Positive Sum: Improving North-South Negotiation*. New Brunswick, NJ: Transaction Press.

Zartman, I. William, and Maureen R. Berman. 1982. *The Practical Negotiator*. New Haven, CT: Yale University Press.

Conference Papers

Regional Regimes: Regional Seas and Shared Freshwater Resources

UNCED. 1991a. *Protection of Oceans, All Kinds of Seas Including Enclosed and Semi-Enclosed Seas, Coastal Areas and the Protection, Rational Use and Development of Their Living Resources: Background Paper.* A/CONF. 151/PC/69 (UNCED PrepCom III).

UNCED. 1991b. *Protection of the Quality and Supply of Freshwater Resources: Application of Integrated Approaches to the Development, Management and Use of Water Resources.* A/CONF. 151/PC/73 (UNCED PrepCom III).

UNEP. 1990a. *Technical Annex to the Report on the State of the Marine Environment.* UNEP Regional Seas Reports and Studies 114/1. Nairobi, Kenya: UNEP.

Protection of the Ozone Layer: The Montreal Protocol

UNEP. 1990b. Open-Ended Working Group of the Parties to the Montreal Protocol. Second Session of the Second Meeting. *Points of Agreement and Disagreement between the Two Studies on the Costs to Developing Countries of Meeting the Objectives of the Montreal Protocol.* UNEP/Ozl.Pro.WG.II(2)/5.

UNEP. 1990c. Open-Ended Working Group of the Parties to the Montreal Protocol. *Report on the First Session of the Third Meeting.* UNEP/Ozl.Pro.WG. III(1)/3.

UNEP. 1990d. Second Meeting of the Parties to the Montreal Protocol on Substances That Deplete the Ozone Layer. *Report of the Executive Director of the United Nations Environment Programme, Secretariat of the Montreal Protocol.* UNEP/Ozl.Pro.2/2/Add.4/Rev.1.

UNEP. 1990e. Second Meeting of the Parties to the Montreal Protocol on Substances That Deplete the Ozone Layer. *Report of the Executive Director of the United Nations Environment Programme, Secretariat for the Vienna Convention and its Montreal Protocol.* UNEP/Ozl.Pro.2/2/Add.5.

UNEP. 1990f. Second Meeting of the Parties to the Montreal Protocol on Substances that Deplete the Ozone Layer. *Report of the Executive Director of the United Nations Environment Programme, Secretariat for the Vienna Convention and its Montreal Protocol.* UNEP/Ozl.Pro.2/2/Add.6 Rev.1.

UNEP. 1990g. Second Meeting of the Parties to the Montreal Protocol on Substances that Deplete the Ozone Layer. *Report of the Executive Director of the United Nations Environment Programme, Secretariat for the Vienna Convention and its Montreal Protocol.* UNEP/Ozl.Pro.2/2/Add.3.

UNEP. 1989a. *Reports of the Ozone Scientific Assessment, Economic and Environmental Effects Panels.* Nairobi, Kenya: UNEP.

UNEP. 1989b. *Report of the Informal Working Group of Experts on Financial Mechanisms for the Implementation of the Montreal Protocol.* UNEP/Ozl. Pro. Mech. 1/Inf.1.

UNEP. 1989c. Open-Ended Working Group of the Parties to the Montreal Protocol. *Final Report of the Second Session of the First Meeting.* UNEP/Ozl.Pro. WG.1(2)4.

UNEP. 1989d. Open-Ended Working Group of the Parties to the Montreal Protocol. *Report of the Third Session of the First Meeting.* UNEP/Ozl.Pro.WG.1 (3)/3.

UNEP. 1989e. Open-Ended Working Group of the Parties to the Montreal Protocol. *Report of the First Session of the Second Meeting.* UNEP/Ozl.Pro.WG.II(1)/7.

UNEP. 1989f. Open-Ended Working Group of the Parties to the Montreal Protocol. First Session of the Second Meeting. *Report of the Legal Drafting Group.* UNEP/Ozl.Pro.WG.II(1)/5.

UNEP. 1986. *Report of the Eighth Session of the Coordinating Committee on the Ozone Layer.* UNEP/CCOL/VIII. Nairobi, Kenya: UNEP.

UNEP. 1985. Conference of Plenipotentiaries on the Protection of the Ozone Layer. *Final Report of the Ad Hoc Working Group of Legal and Technical Experts for the Elaboration of a Global Framework Convention for the Protection of the Ozone Layer.* UNEP/IG. 53/4.

UNEP. 1984a. Ad Hoc Working Group of Legal and Technical Experts for the Elaboration of a Global Framework Convention for the Protection of the Ozone Layer. *Second Revised Draft Protocol Concerning Measures to Control, Limit and Reduce the Emissions of Chlorofluorocarbons (CFCS) for the Protection of the Ozone Layer.* UNEP/WG. 94/12.

UNEP. 1984b. Ad Hoc Working Group of Legal and Technical Experts for the Elaboration of a Global Framework Convention for the Protection of the Ozone Layer. Fourth session. *Summary of Comments Received from Governments on the Fourth Draft Convention and Second Revised Draft Protocol.* UNEP/WG. 110/2.

UNEP. 1983. Ad Hoc Working Group of Legal and Technical Experts for the Elaboration of a Global Framework Convention for the Protection of the Ozone

Layer. *Revised Draft Protocol Concerning Measures to Control, Limit and Reduce the Emissions of Chlorofluorocarbons (CFCS) for the Protection of the Ozone Layer*. UNEP/WG. 94/9.

UNEP/ICSU/WMO. 1986. *Report of the International Conference on the Assessment of the Role of Carbon Dioxide and of Other Greenhouse Gases in Climate Variations and Associated Impacts*. WMO Report no. 661. Geneva: WMO.

UNEP/Ozl. 1990a. Montreal Protocol Working Group II, (2)/27 February–5 March.

UNEP/Ozl. 1990b. Montreal Protocol Working Group III (2)/3 Annex III p. 12, June.

The Basel Convention on Hazardous Wastes

UNCED. 1992. A/CONF. 151/PC/100/Add.24/Annex I (UNCED PrepCom IV).

UNCED. 1991c. *Environmentally Sound Management of Toxic Chemicals: Progress Report of the Secretary-General of the Conference*. A/CONF.151/PC/35 (UNCED PrepCom II).

UNCED. 1991d. *The International Economy and Environment and Development*. A/CONF.151/PC/47 (UNCED PrepCom III).

UNCED. 1991e. *Progress Report on Financial Resources*. A/CONF.151/PC/51 (UNCED PrepCom III).

UNCED. 1991f. *Report on Transfer of Technology*. A/CONF.151/PC/53 (UNCED PrepCom III).

UNCED. 1991g. *Prevention of Illegal International Traffic in Toxic and Dangerous Products and Wastes*. A/CONF. 151/PC/88 (UNCED PrepCom III).

UNCED. 1991h. *Protection of the Atmosphere: Sectoral Issues*. A/CONF. 151/PC/60 (UNCED PrepCom III).

UNEP. 1991h. *Recommendations on International Strategy and Action Programme Including Technical Guidelines for the Environmentally Sound Management of Hazardous Wastes of the Ad Hoc Meeting of Government-Designated Experts*.

UNEP. 1989a. *London Guidelines for the Exchange of Information on Chemicals in International Trade, Amended*.

UNEP. 1989b. Ad Hoc Working Group of Legal and Technical Experts with a Mandate to Prepare a Global Convention on the Control of Transboundary Movements of Hazardous Wastes. *Proposals by the Executive Director for Consideration by the Ad Hoc Working Group at Its Fourth Session*. UNEP/WG.190/3.

UNEP. 1989c. Ad Hoc Working Group of Legal and Technical Experts with a Mandate to Prepare a Global Convention on the Control of Transboundary Movements of Hazardous Wastes. *Report of the Ad Hoc Working Group on the Work of Its Fourth Session*. UNEP/WG.190/4.

UNEP. 1989d. *Informal Negotiations on Hazardous Wastes: Points Identified by the Executive Director for Further Consideration at the Informal Negotiations on Hazardous Wastes.*

UNEP. 1989e. Conference of Plenipotentiaries on the Global Convention on the Control of Transboundary Movements of Hazardous Wastes. *Final Report of the Ad Hoc Working Group of Legal and Technical Experts with a Mandate to Prepare a Global Convention on the Control of Transboundary Movements of Hazardous Wastes.* UNEP/IG. 80/4.

UNEP. 1988. Ad Hoc Working Group of Legal and Technical Experts with a Mandate to Prepare a Global Convention on the Control of Transboundary Movements of Hazardous Wastes. *Report of the Ad Hoc Working Group on the Work of Its Third Session.* UNEP/WG.189/3.

Conservation of Biological Diversity

UNCED. 1991i. *Conservation of Biological Diversity.* A/CONF. 151/PC/28 (UNCED PrepCom II).

UNCED. 1991j. *Environmentally Sound Management of Biotechnology.* A/CONF. 151/PC/29 (UNCED PrepCom II).

UNCED. 1991k. *Conservation of Biological Diversity: Background and Issues.* A/CONF. 151/PC/66 (UNCED PrepCom III).

UNEP. 1993. Intergovernmental Committee on the Convention on Biological Diversity. First session. *Issues before the Intergovernmental Committee on the Convention on Biological Diversity. Note of the Executive Director.* UNEP/CBD/IC/1/3.

UNEP. 1992a. Intergovernmental Negotiating Committee for a Convention on Biological Diversity. Sixth negotiating session/Fourth session of INC. *"Agreed Incremental Costs" for the purpose of the Convention on Biological Diversity.* UNEP/Bio.Div/n.6-INC.4/3.

UNEP. 1992b. Intergovernmental Negotiating Committee for a Convention on Biological Diversity. *Report of the Intergovernmental Negotiating Committee for a Convention on Biological Diversity on the Work of Its Sixth Negotiating Session/Fourth session of INC.* UNEP/Bio.Div/N6-INC.4/4.

UNEP. 1992c. Intergovernmental Negotiating Committee for a Convention on Biological Diversity. Seventh negotiating session/Fifth session of INC. *Informal Note by the Chairman of the INC and the Executive Director of UNEP Regarding Possible Compromise Formulations for the Fifth Revised Draft Convention on Biological Diversity.*

UNEP. 1992d. Intergovernmental Negotiating Committee for a Convention on Biological Diversity. Seventh negotiating session/Fifth session of INC. *Fifth Revised Draft Convention on Biological Diversity. Explanatory Note.* UNEP/Bio.Div/N7-INC.5/2.

UNEP. 1992e. First Meeting of the Conference of the Parties to the Basel Convention on the Control of Transboundary Movements of Hazardous Wastes and Their Disposal. *Report of the First Meeting of the Conference of the Parties to the Basel Convention.* UNEP/CHW.1/24.

UNEP. 1991i. Ad Hoc Working Group of Legal and Technical Experts on Biological Diversity. *Report on Its Second Session.* UNEP/Bio.Div/WG.2/2/5.

UNEP. 1991j. Ad Hoc Working Group of Legal and Technical Experts on Biological Diversity. Third session. *Description of Transferable Technologies Relevant to Conservation of Biological Diversity and Its Sustainable Use. Note By the Secretariat.* UNEP/Bio.Div/WG.2/3/10.

UNEP. 1991k. Ad Hoc Working Group of Legal and Technical Experts on Biological Diversity. Third session. *A Preliminary Note on the Concepts Outlined in Some of the Key Terms and Phrases Used in the Draft Articles.* UNEP/Bio.Div/WG.2/3/6.

UNEP. 1991l. Ad Hoc Working Group of Legal and Technical Experts on Biological Diversity. Third session. *Note to Facilitate Understanding of Issues Contained in Articles under Consideration by Sub-Working Group II.* UNEP/Bio.Div/WG.2/3/7.

UNEP. 1991m. Ad Hoc Working Group of Legal and Technical Experts on Biological Diversity. Third session. *A Note on Options for a Financial Mechanism to Meet the Requirements of a Convention on Biological Diversity.* UNEP/Bio.Div/WG.2/3/4.

UNEP. 1991n. Ad Hoc Working Group of Legal and Technical Experts on Biological Diversity. Third session. *A Note in Treaty Language Containing the Different Options for a Financial Mechanism Based on Solutions Adopted in Other Conventions and Other Multilateral Financial Mechanisms.* UNEP/Bio.Div/WG.2/3/8.

UNEP. 1991o. Ad Hoc Working Group of Legal and Technical Experts on Biological Diversity. Third session. *Note on the Legal Instruments in Existence Relevant to Access to Biological Diversity Outside Areas of National Jurisdiction. Summary.* UNEP/Bio.Div/WG.2/3/9.

UNEP. 1991p. Ad Hoc Working Group of Legal and Technical Experts on Biological Diversity. Third session. *A Note Clarifying a Clearing House Mechanism on Transfer of Technology and Technical Co-operation.* UNEP/Bio.Div/WG.2/3/5.

UNEP. 1991q. Intergovernmental Negotiating Committee for a Convention on Biological Diversity. Fourth session. *Note by the Executive Director.* UNEP/Bio.Div/N.6-INC.4/3.

UNEP. 1991r. Intergovernmental Negotiating Committee for a Convention on Biological Diversity. Fourth session. *Interpretation of Phrases "Adequate, New and Additional," "New and Additional" and "Adequate and Additional" Financial Resources.* UNEP/Bio.Div/N.6-INC.4/4.

UNEP. 1991s. Intergovernmental Negotiating Committee for a Convention on Biological Diversity. Fifth negotiating session/Third session of INC. *Interpretation of the Words and Phrases "Fair and Favourable," "Fair and Most Favourable," "Equitable," "Preferential and Non-Commercial," "Preferential," "Non-Commercial" and "Concessional."* UNEP/Bio.Div/N.5-INC.3/3.

UNEP. 1991t. Intergovernmental Negotiating Committee for a Convention on Biological Diversity. Sixth negotiating session/Fourth session of INC. *Fourth Revised Draft Convention on Biological Diversity. Explanatory Note.* UNEP/Bio.Div/N6-INC.4/2.

UNEP. 1990h. *Ad Hoc Working Group of Legal and Technical Experts on Biological Diversity: Report on the Work of Its First Session.* UNEP/Bio.Div/WG.2/1/4.

UNEP. 1990i. *Ad Hoc Working Group of Experts on Biological Diversity. Report on the Work of Its Second Session in Preparation for a Legal Instrument on Biological Diversity of the Planet.* UNEP/Bio.Div.2/3.

UNEP. 1990j. Ad Hoc Working Group of Experts on Biological Diversity. Third session. *Biological Diversity: Global Conservation. Needs and Costs Executive Summary.* UNEP/Bio.Div.3/3.

UNEP. 1990k. Ad Hoc Working Group of Experts on Biological Diversity. Third session. *Ongoing Discussions on Intellectual Property Rights in Uruguay Round of GATT Negotiations.* UNEP/Bio.Div.3/8.

UNEP. 1990l. Ad Hoc Working Group of Experts on Biological Diversity. Third session. *Ongoing Discussions on Intellectual Property Rights in UPOV, WIPO and GATT as They Relate to Access to Genetic Resources.* UNEP/Bio.Div.3/11.

UNEP. 1990m. Ad Hoc Working Group of Experts on Biological Diversity. Third session. *Relationship between Intellectual Property Rights and Access to Genetic Resources and Biotechnology. A Study Prepared for UNEP.* UNEP/Bio.Div.3/Inf. 4.

UNEP. 1990n. Ad Hoc Working Group of Experts on Biological Diversity. Third session. *Relevant Existing Legal Instruments, Programmes and Action Plans on Biological Diversity.* UNEP/Bio. Div.3/Inf. 6.

UNEP. 1990o. Ad Hoc Working Group of Experts on Biological Diversity. *Report of the Ad Hoc Working Group on the Work of Its Third Session in Preparation for a Legal Instrument on Biological Diversity of the Planet.* UNEP/Bio.Div.3/12.

UNEP. 1990p. Ad Hoc Working Group of Experts on Biological Diversity. Third session. *Relationship between Intellectual Property Rights and Access to Genetic Resources and Biotechnology.* UNEP/Bio.Div/3/6.

UNEP. 1990q. Ad Hoc Working Group of Experts on Biological Diversity. Third session. *Biotechnology: Concepts and Issues for Consideration in Preparation of a Framework Legal Instrument for the Conservation of Biological Diversity.* UNEP/Bio.Div.3/7.

UNEP. 1988. Ad Hoc Working Group of Experts on Biological Diversity. *Report of the Ad Hoc Group of Experts to the Executive Director on Governing Council Decision 14/26*. UNEP/Bio. Div.1/Inf.1.

UNEP. 1975. *Report of the Governing Council on the Work of Its Third Session*. Nairobi, Kenya: UNEP.

UNEP. 1974. *Introductory Report by the Executive Director to the Second Session of the Governing Council*. UNEP/GC/14. Nairobi, Kenya: UNEP.

UNEP Series on Environmental Law Guidelines and Principles

UNEP. 1987. *Environmental Law: Guidelines and Principles (8). Environmentally Sound Management of Hazardous Wastes.*

UNEP. 1987. *Environmental Law: Guidelines and Principles (9). Environmental Impact Assessment.*

UNEP. 1987. *Environmental Law: Guidelines and Principles (10). Exchange of Information on Chemicals in International Trade.*

UNEP. 1985. *Environmental Law: Guidelines and Principles (7). Marine Pollution from Land-Based Sources.*

UNEP. 1984. *Environmental Law: Guidelines and Principles (6). Banned and Severely Restricted Chemicals.*

UNEP. 1982. *Environmental Law: Guidelines and Principles (4). Offshore Mining and Drilling.*

UNEP. 1980. *Environmental Law: Guidelines and Principles (3). Weather Modification.*

UNEP. 1978. *Environmental Law: Guidelines and Principles (2). Shared Natural Resources.*

Index